D U S T

D U S T

The Inside Story of Its Role in the September 11th Aftermath

Paul J. Lioy

ROWMAN & LITTLEFIELD PUBLISHERS, INC.
Lanham • Boulder • New York • Toronto • Plymouth, UK

Published by Rowman & Littlefield Publishers, Inc.
A wholly owned subsidary of The Rowman & Littlefield Publishing Group, Inc.
4501 Forbes Boulevard, Suite 200, Lanham, Maryland 20706
http://www.rowmanlittlefield.com

Estover Road, Plymouth PL6 7PY, United Kingdom

British Library Cataloguing in Publication Information Available

Library of Congress Cataloging-in-Publication Data

The hardback edition of this book was previously cataloged by the Library of
Congress as follows:

Lioy, Paul J.
  Dust : the inside story of its role in the September 11th aftermath / Paul J. Lioy.
    p. cm.
  Includes bibliographical references and index.
  1. Dust—New York (State)—New York—Environmental aspects. 2. Dust—
Analysis. 3. Dust—Health aspects. 4. September 11 Terrorist Attacks,
2001—Health aspects. 5. Air—Pollution—Measurement. 6. World Trade
Center Site (New York, N.Y.) 7. Pulmonary toxicology. I. Title.
  TD884.5.L568 2010
  363.738—dc22   2009031559

ISBN 978-1-4422-0148-4 (cloth : alk. paper)
ISBN 978-1-4422-0149-1 (pbk. : alk. paper)
ISBN 978-1-4422-0150-7 (electronic)

To Alexander and Samuel Lioy
"Live long and prosper." —Spock in *Star Trek*

To Jeanie and Jason, and Laura,
"Thanks for the memories." —Bob Hope

CONTENTS

FOREWORD

The rationale behind the establishment of the Environmental and Occupational Health Science Institute (EOHSI) began to evolve during the 1980s—a time that saw multiple environmental crises affecting our state along with increasing pressure to create more jobs. I have never believed that these two goals have to be in conflict. From my early efforts to preserve Sunfish Pond in Northwest New Jersey, to my work on complex environmental issues, such as dioxin and radon, it became clear that there needed to be scientific data to supplement and support our environmental policies. In 1986, I met with leaders from the scientific and academic communities to create an institute to attract the top scientists in various fields and disciplines who would provide in-depth scientific research and analysis without the intrusion of politics. Our goal was to ensure that our environmental policies had sound scientific backing and to support clean jobs for our future. It was clear that New Jersey needed to attract the brightest minds in order to cultivate the best jobs and attract the best industries to our state.

The unique landscape of New Jersey with its rural, suburban, coastal, and urban areas has led to the need for unique and sensitive environmental policies. However, with increased business and residential development occurring throughout the state, the usage of accurate, scientific data to form environmental policy became mandatory. The EOHSI has been a thought leader since its inception and provided legislators and officials with data that has helped us understand and protect our environment. The work performed by the institute and its increased recognition in the scientific community has helped form relationships that have allowed the institute to expand and become a leader in scientific analysis and research. That recognition gave the institute the ability to network with other scientists and agencies to confront serious environmental questions and to add consistent analysis to the public debate on matters of importance.

The work carried out by the institute immediately following the September 11, 2001, terror attack on the World Trade Center shows how important this organization is to public health planning and policy. In the days after the attack, countless volunteers streamed in to lend a hand with the rescue operations. At the same time fears began to surface regarding the toxic environment created by the collapse of the Twin Towers. Dr. Paul Lioy and his EOHSI team rushed to Ground Zero to collect valuable exposure samples and data. It was this response— remarkable for its speed and professionalism—that enabled the EOHSI team to collect unique, time-sensitive samples from key areas of Ground Zero. And it is this rare data and analysis that forms the basis of Dr. Lioy's book.

I am proud of the institute. The work they have done over the years has been vital to protecting the Garden State and our neighbors. It was just such work that I envisioned at the institute's beginning. Thanks to their hard work, the EOHSI's findings have shown that the supply and usage of protective gear and understanding the dangers of exposure are vital for future disaster recovery and city planning efforts. I thank Dr. Lioy and his team for their hard work, and I hope that their work serves as a foundation to develop better environmental policy for future generations.

Governor Tom Kean

PROLOGUE

Within twenty four hours of the September 11 attack on the World Trade Center (WTC) in New York City, representatives from several governmental agencies asked me about the dust that was released during the collapse of each tower. The thick gray and fluffy dust seemed to be everywhere, settling on all of the animate and inanimate objects in its path. It covered the skin and clothes of many of those who had survived but who had been trapped in harm's way. You could see it being resuspended in the air after official vehicles drove through Manhattan. What was in that dust and its companion plume of smoke that was moving across Brooklyn and out to sea? At that time, I didn't know the answer to this question.

A few days later, colleagues from the Environmental and Occupational Health Sciences Institute (EOHSI), New Jersey, and I were granted access to the area that had quickly become known as Ground Zero. Since I wanted to know what was in the dust, a colleague and I gathered samples for chemical and physical analyses. As I engaged in this task, I couldn't help experiencing thoughts and emotions that were far from the usual realm of data collection and analytical studies that I had completed or directed throughout my scientific career.

Little did I realize that the WTC dust would take on a life of its own and would affect my professional career for years to come. I also became painfully aware of the seriousness of the security issues facing America. I think many people of my generation have recognized the life-altering nature of the event—some more than others, and not all in the same way. The story about the WTC dust and me would take many turns, none of which were anticipated during the time when we took samples outdoors and, later, indoors in Lower Manhattan. In 2008, the WTC dust was even noted in the movie *American Carol*, as a poignant moment when George Washington is about to open the door from St. Paul's Chapel in Lower Manhattan, and you see Ground Zero.

The scientific issues caused by the events of 9/11 and their aftermath have been reported in the technical literature by my colleagues and me, and others, in scholarly journals. However, the time course of the science, the analysis of the results, and the lessons learned (or not) have only been developed in a fragmented fashion and not clearly described for the general public. Furthermore, for some reason, people keep analyzing WTC dust samples. These analyses have been completed by a number of scientists after individuals had asked me for samples, or after stored dust samples were discovered or WTC dust was found and removed from previously stored personal effects. In any case, there is continued interest. In fact, on the seventh anniversary, I still received a request for a portion of the WTC dust samples. However, because of the age of the samples and possible transformation of the material, I have decided to only honor requests from individuals who want to use them primarily in educational programs or in research as histori-cal artifacts. Since all the major constituents have been identified, any new materials that would be found in samples would not change my overall conclusions about exposures to the WTC dust and its role in the aftermath of 9/11.

Since I was not at the World Trade Center on September 11, it would be impossible for me to accurately describe the scene; however, to pro-vide a setting for the WTC dust, which is the subject of this story, I will share some personal background and a brief discussion of the events of 9/11. These will give you the context for why I reacted the way I did to the events of that day.

On the 11th, just about everyone had television or access to a televi-sion, which brought much of the horror continuously into our homes— as I usually say to people, everything but the smell and the background noise. In contrast, when I was young, the graphic images of serious and deadly past events were not so readily available to the public; a perspective that some of you may not have considered in the years following the attack on the WTC. In fact, the information and insights I gathered on various aspects of World War II and other events farther back in time, like the Civil War, could mostly be found in books, pictures (World War II), and lithographs (Civil War). During World War II, newsreels and other limited documentaries were shown in movie theaters. As a boy, identifying and understanding true heroism and the trauma that took place during World War II was found by watching and rewatching *Victory at Sea* with my father, who never said much about the war.

I never understood the lack of discussion about World War II at family gatherings on weekends and holidays, which is in stark contrast to the discussion about the attack on the World Trade Center. In fact, I ignored the deeper psychological effects of the battles and survival during World War II until I read Tom Brokaw's *The Greatest Generation*, which was just after I turned fifty-one! I was so moved by the content of the book that I wrote Brokaw a letter, and I believe the following excerpts from that letter are relevant to how we need to view the September 11 aftermath in the future:

I read the *Greatest Generation* last year and recently completed the *Greatest Generation Speaks*. They are an exceptional two volume set (now three) because they finally put the lives of my parents, my in-laws, aunts and uncles, and close family friends into a meaningful perspective. Each of these people had strengths and weaknesses, but one overriding consistency was that none of the couples were ever divorced. This must say something about the resilience and bonds that were formed among these people between 1941 and 1945. . . . It is a pity, however, that most of the men in the group died between one and five years prior to the release of these books. I believe each would have appreciated your compassion for their struggles and your understanding of their efforts.

Personally, I have always had a major interest in the events and conditions associated with World War II. One sobering experience for me was a visit to the Arizona Monument in Hawaii, and then attempting to recreate images of the chaos on that "Day of Infamy," December 7, 1941. This group of men and women in my family never talked openly about that day or the war years, which underscores a theme discussed in both your books. . . .

Now I have a richer and deeper insight into why they just wanted to move on with their lives. On the downside I cannot reach out to most of the above and say, "Thank you for your *resilience* in the face of grave danger to our way of life."

Baby boomers would have benefited from these or similar books while growing up, but that is water under the bridge. Furthermore, the movie *Best Years of Our Lives* was released ten years too early and *Saving Private Ryan* and *Band of Brothers* were released over thirty years too late to help my generation understand the life and times of their parents from 1941 to 1945. Therefore, from the standpoint of September 11, in addition to the books and information about the horror itself, it is also important for my son's generation (Jason is now thirty-six) and future young Americans to also learn about *the science of the dust and the types of exposures* that occurred after our day of infamy.

While there is no question that my personal interest in World War II influenced my perceptions about the attack on the World Trade Center and the need for acquiring information, it was my scientific training that was essential in finding the correct role. My professional career in environmental science started in 1975 after graduating from Rutgers University with a Ph.D. This would have never happened if my wife, Jeanie, had not pointed out that the environment was a "hot new field" and could probably use my skill sets. Rutgers also gave me the foundation for completing interdisciplinary research and policy activities within the environmental field during the years ahead.

In 1978, I joined the NYU Institute of Environmental Medicine and had the great fortune of working with and learning from some of the pioneers in environmental health sciences. I also met two colleagues and friends with whom I have shared many experiences for over twenty years: Morton Lippmann, who will appear later in the book, and Bernard Goldstein, who in 1985 brought me to EOHSI (NJ) at the Robert Wood Johnson Medical School of the University of Medicine and Dentistry of NJ, and Rutgers University, where I still work. At EOHSI I became part of a new scientific field of study now called exposure science. It will be mentioned throughout the book because of its direct relationship to WTC dust.

As defined by the *Journal of Exposure Science and Environmental Epidemiology*, exposure science is "the study of human contact with chemical, physical, and biological agents occurring in their environments, and advancement of knowledge of the mechanisms and dynamics of events either causing or preventing health outcomes." In layman's terms, it is the how, what, where, and why we come or can come in *contact* with toxic materials in the world around us. This contact is caused by either our own actions or the actions of others, and in certain situations the exposures can lead to illness or disease. Thus, the basic concepts in exposure are easy to discuss since we all have many types of exposures every day, such as contact with chemical, biological (e.g., bacteria), and physical (e.g., radon) agents. However, measuring them and understanding the significance of such contacts and exposures to human health requires both basic and applied research and analysis, which can span from basic chemical analysis to human behavior assessment. Thus, the work that I was involved with over the years actually became part of the "birth of a scientific field" and an important component of the WTC aftermath.

I am an American, a scientist, and a second-generation Italian American born to Nick and Jean. I grew up in Passaic, New Jersey. My hometown is located about fifteen miles west of what I still maintain is one of the greatest cities in the world, New York City. On April 11, 2005, my son Jason and his wife Laura's son Alexander was born. Obviously, this was a very special day for me and my family. However, for the rest of my life, the 11th of any month holds darker feelings because of the events of September 11, 2001. Alexander and now his brother Samuel are part of the post-9/11 generation, and they will only learn about 9/11 from the stories told about the events of that day and beyond. The long-term effects on my grandsons and many other children are still to be played out, and those will be colored by the way we tell the story of 9/11. I hope that they develop personal resiliency and a better sense of what is important to the security of our country, and embrace the quote by President Kennedy: "Ask not what your country can do for you, but what you can do for your country."

Dust reflects my personal conclusions and professional recommendations as a result of having assessed the environmental and occupational exposures in the aftermath of that horrific day in American history. It

could not have been written before now because it took time to collect
and analyze the scientific data that resulted from the collapse of the
WTC. Then it took time for me to fully understand the consequences
of the policy decisions we made then, and put the story into a form that
could be understood by those who are not in science or only had limited
time to focus on the details. The book also focuses on what we did and
did not learn in the WTC aftermath and how we can best use our September 11 experience to deal with exposures during catastrophic events
in the future.

ACKNOWLEDGMENTS

Many people contributed to parts of this story on the World Trade Center dust. I will name many but probably do not remember all, so my apologies to those individuals. First are my WTC dust collaborators. I thank them for their efforts and the many discussions we have had over the years. Included are my longtime colleagues and friends Drs. Clifford Weisel and Panos Georgopoulos, as well as Drs. Daniel Vallero, John Offenberg, Brian Buckley, Steve Eisenreich, Charles Weschler, Robert Hale, and Barbara Turpin. I also thank Jim Millette, a collaborator who provided the many micrographs of the dust and interesting discussions on forensic sciences. Dr. Gary Foley, of the Environmental Protection Agency (EPA), had the vision to understand the value of completing the plume model. I thank the "E-team" at the Environmental and Occupational Health Sciences Institute and of Robert Wood Johnson Medical School–University of Medicine and Dentistry of New Jersey (EOHSI–UMDNJ), Drs. Michael Gallo and Michael Gochfeld, with whom I have worked on many aspects of WTC issues; and Drs. Mark Robson and Howard Kipen, for many conversations and their personal efforts following the 9/11 attack. I thank members of the National Institute of Environmental Health Sciences Centers; Drs. Phil Landrigan and Mary Wolff of Mount Sinai; Drs. Morton Lippmann and George Thurston of New York University (NYU) Medical Center; Dr. Lung Chi Chen, my main collaborator from NYU, who graciously provided photos presented in the book; Dr. David Prezant, of the Fire Department of New York, for his insights and lasting friendship, one forged because of this tragedy; Dr. Edo Pellizzari, a friend and colleague with whom I have shared many scientific adventures, and many hours of discussion about science and the world around it; Richard Canas, director of the Office of Homeland Security and Preparedness for the State of New Jersey, for providing me with many insights about his work and ways my research interests can lead to useful applications within emergency preparedness and response.

A special thank-you to Dr. Hugh Tilson, editor of the journal *Environmental Health Perspectives*, for giving permission to use many of the pictures and text that we published in our 2002 article on the WTC dust. I also appreciate the use of the dust movement figure in southern Manhattan with the permission of the New York City Police Department.

From EPA Region II, Dr. Mark Maddaloni added some important details about EPA's efforts, and Kathy Callahan was persistent in getting things done in downtown Manhattan in the face of many odds. Dr. Matt Lorber, of EPA in Washington, offered valuable insights on risk assessment and future needs.

I thank Dr. Alison Geyh of Johns Hopkins for her research and interesting conversation about the worker issues during the recovery period at Ground Zero. I am grateful to the members of the WTC Expert Technical Panel, with special thanks to Drs. Paul Gilman, Mort Lippmann, Joseph Picciano, and Greg Meeker. Thanks also to all my staff, including Teresa Boutillette and Linda Everett, for sharpening EOHSI modeling figures, all other division staff and graduate students at EOHSI who participated in the sampling and analysis; many are named as authors on individual manuscripts.

Thank you also to my son, Jason, and to our friends Gene and Ruth Marino, and Buzz and Jane Waltman, whose encouragement led to my finally putting hands to the keyboard for this book, and for the comment "Write it for your grandsons." To my friend Bob Mulcahy, who was one of the first to say, "Write the book," and to Anthony DePalma for interesting conversations and very thoughtful questions. To Laura Lioy, my daughter-in-law, for her help in organizing and cleaning up my photos and insights on the cover design; and to my agent, Joan Parker, for all her efforts on this project. I also want to thank my publisher for taking on this book, especially Suzanne I. Staszak-Silva, and the staff at Rowman & Littlefield, especially Julia Loy. Thanks also to the former governor of New Jersey, Thomas H. Kean, for his vision in providing the resources that supported the creation of EOHSI and for graciously agreeing to write the foreword for this book, and to his colleagues Anthony Cicatiello and Matthew Deluca.

Finally, to my wife, Professor Mary Jean (Jeanie) Yonone-Lioy, a special thanks for all her intellect, patience, and advice over these many years. For taking the time to comment on the initial manuscript, keeping me focused on the story, and for her efforts in educating many responders about anthrax and other biological WMDs.

1

WHEN LIFE CHANGED

The *U.S. News & World Report* cover story for the week of September 3, 2001, was about Americans finding ways to achieve happiness, and on the cover was a big yellow happy face. The story line was "How to Make Yourself Happy: Cheer up! New science says you can do lots more to inject real joy into your life." We were also still recovering from a presidential election where two of the biggest issues were a "lockbox" for Social Security and the economic recovery from the dot-com mess. Plus, we learned more than we ever wanted to know about "hanging chads." We had also just gone through the Y2K scare—I remember watching the 2000 New Year's celebration around the world on TV, feeling relieved that nothing bad had happened. The New York Yankees had beaten the Mets not quite a year before in major league baseball's first Subway Series since the 1950s. Clearly, the "winds of war" were not foremost on our minds, as individuals or collectively as a nation.

On September 11, I got up late and, as I remember it, suddenly my wife, Jeanie, came running into the house, turned on the TV, and her eyes were glued to the set. We saw the smoke coming from the North Tower of the World Trade Center and concerns being raised by the commentators. At first we thought that this might have been a horrible small plane accident, but we both had some vague thoughts about maybe something worse. Then the unthinkable happened at 9:03 AM. The second plane flew straight into the South Tower, while we were watching! In an instant we knew that this was no longer an accident. But what was coming next? The next half hour is well-documented history. As Mike Gochfeld, a colleague from EOHSI and part of this story, would say, "9/11 was a defining moment on various levels from international to national to regional, to New York to New Jersey, and even to our own professional activities" (Mike Gochfeld, pers. comm.).

Two more planes were hijacked and more Americans were killed when Flight 77 flew into the Pentagon in Washington, DC. My colleague Robert Snyder, who was on his way to a meeting on Acute Exposure Guidelines (AEGLs) for highly toxic substances, had just deplaned at Reagan National Airport when he heard the explosion and saw the smoke rising from the Pentagon.

During that time, like many folks, we first began to think about family. Where was our son, Jason? Fortunately, he was at work in Pennsylvania. However, where was our future daughter-in-law, Laura? She was in New York City. We found out she was safe but was trying to find a way out of the city. Jason said this was "a horrible day," wondering "how could we let that happen here?" He eventually talked to Laura, and then we did. She got back to Pennsylvania by train the next morning. Laura was in Times Square at the time of the attack and saw the WTC burning. She tried to leave using the Lincoln Tunnel, but after the second tower was hit, she could not get out of the city and began to lose her cell phone contacts. Because of her location during the aftermath, she did not see the dust clouds. On that day Laura tried to help by donating blood. She had nothing else to do, and later that day she ate dinner in what she described as a "ghost town."

I am sure that these moments of uncertainty, confusion, and more were experienced by many individuals and families that day, but for thousands of people, a dreadful day would soon turn into a personal nightmare. At the time, we were also looking at one part of what would become the WTC pollution aftermath, the smoke that was being emitted from the jet fuel fires.

In contrast to the attack on Pearl Harbor where details were vague or totally missing, on September 11, 2001, media access was different:

we were seeing the events as they unfolded right before our collective eyes. In fact, there are books that detail just about every minute of that day—for example, *What We Saw* by CBS News. Everything was in front of us, which even at a distance of twenty miles to the southwest actually helped me begin to frame the health science issues that developed during the initial hours of the aftermath.

The horror of the event was becoming frightfully clear, as the air space was shut down. No more attacks occurred, but there was this plane in Pennsylvania, Flight 93. It was now 9:45 AM. The fires blazed above the towers, and the graphic displays on the TV were nonstop. At one point the screens were beginning to repeat events, as the fires burned, and the speculation began about who did it. There were no fighter jets flying overhead. The reality then began to really sink in: we had been attacked with our own planes, commercial airliners—killing and injuring innocent people. This was a true act of war, a sneak attack on two major hubs of commerce and power, New York and Washington, DC. Then, as many will remember, the next horrific moment happened at 9:50 AM. The South Tower of the WTC suddenly collapsed. It was not an ordinary collapse. The building went down like a piston and began to quickly disintegrate into a dust cloud that started moving through Lower Manhattan, as shown in figure 1.1. It was both horrific and symbolic since the world of many people in the NYC metropolitan area began to disintegrate with the collapse. More people would die, and many traumas and injuries would be experienced by those who remained behind in the years ahead. A "blizzard of white dust," as noted in a 2003 report from the Government Accounting Office, enveloped the area, yielding a grim reminder of what had just happened.

The collapse of the tower and the "blizzard" of white dust released into Lower Manhattan was the moment that personalized my efforts to get involved with the aftermath. It led directly to my group's efforts to complete many scientific analyses on dust samples and our attempts to understand human exposures to the dust and smoke during the aftermath. The scientific articles that we published, which are the basis for this story, are listed in the bibliography, as well as other citations that are referenced throughout the book. These materials provide information on details of scientific observations that we made and are integrated into the individual chapters of this book. By including the highlights of the

Figure 1.1. The movement of WTC dust throughout southern Manhattan.

results and the conclusions we made in some of the papers cited in this book, I will give you an idea about what and how we were trying to deal with the material known as the WTC dust.

A few minutes past 10:00 AM, Flight 93 crashed in Pennsylvania. What was going to happen next? As we watched the TV, the smoke and dust from the South Tower collapse had moved deeper into Lower Manhattan. At 10:29, the North Tower collapsed, and the second wave of dust and smoke billowed out through southern Manhattan. Was this the end? No one knew. However, in southern Manhattan the dust levels had increased significantly in volume and area of coverage. A satellite picture taken of the area on September 13 (figure 1.2), reviewed and analyzed by the EPA, marked the general extent of the dust deposition on the ground and on many buildings. The actual boundary of the dust deposited was never completely defined, but it would have included spaces north of Canal Street and into the western boarder of Brooklyn, on the East River.

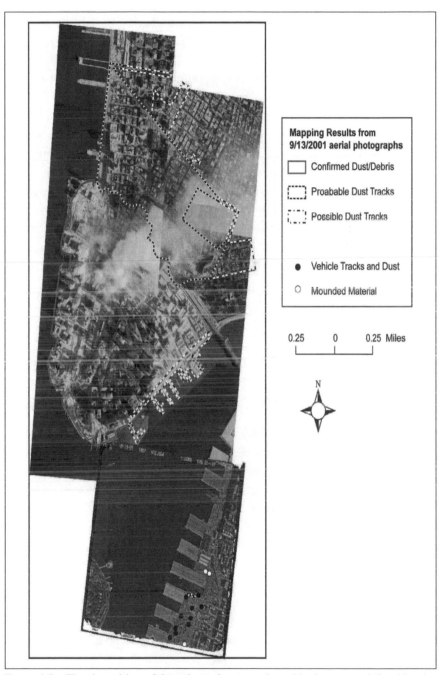

Figure 1.2. The deposition of dust throughout southern Manhattan as defined by the U.S. EPA. Used with permission of the U.S. EPA, Region II.

A state of emergency was declared in New York City and by 1:30 PM across the country. My dean at Robert Wood Johnson Medical School (UMDNJ) determined that all nonessential personnel should go home. I went to my office to see how the trauma from the event was affecting faculty and staff who had remained in the office.

In the metropolitan area, the tragedy was beginning to unfold, with initial estimates of six thousand to ten thousand people dead or dying. Eventually the number was reduced to about three thousand souls lost because of the multiple attacks. Within my personal circle, my neighbor lost his son-in-law in one of the towers, and my gardener lost his niece on Flight 93. In NYC, about twenty thousand families were initially displaced from their downtown homes, and many were still displaced in October. As my colleague Mike Gochfeld would simply but elegantly say, "The sense of security was shattered." It was truly the "Day of Infamy" in my lifetime, affecting everyone.

THE AFTERMATH

The area most directly impacted by the collapse of each tower was Lower Manhattan. Figure 1.3 is a map of this area of New York City that also notes the locations where asbestos samples were taken.

On the evening of the 11th, I began to think about the potential for significant *acute* exposures to dust for many segments of the population who were in harm's way that day. My concern would eventually prove to be correct. However, my thoughts appeared to be at odds with the way other scientists started to examine the environmental and occupational health issues. Some experts seemed to be focused on the measurement of materials that usually cause long-term health effects, including asbestos, which, as shown in figure 1.3, was monitored during the initial days post-9/11. However, in contrast to comments by others, I cannot criticize their initial reactions and response. It was such an overwhelming disaster, and we had no experience with the magnitude and complexity of the WTC aftermath—a major point lost by many in the days and years that followed. Science had to catch up, but there was little time available to do so. For some aspects of the event, there was no time at all. This book will try to put the role of the WTC dust into its proper place. However, this is done

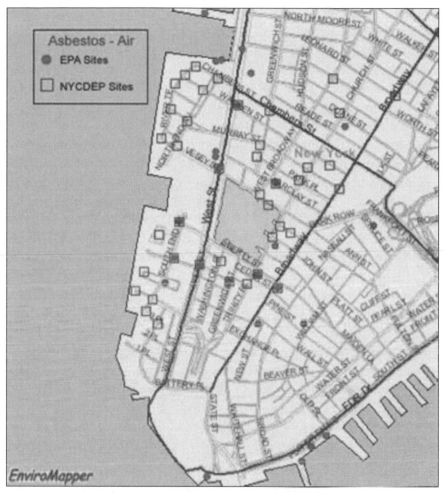

Figure 1.3. Map of Lower Manhattan and the location of EPA and NYC asbestos monitoring sites for the period immediately after the attack on the WTC. Used with permission of the U.S. EPA, Region II.

with my sincere appreciation of other scientists, physicians, and engineers who were also doing their best to cope with the event and its aftermath.

Many criticisms have been directed at the overall response, some justified, many not justified, and some that were and still are misdirected. It is a pity that as a society we just did not get it or take the time to learn. The event was unprecedented, and we just were not prepared to respond. However, several lessons needed to be learned, and I am not sure we have learned them yet. I hope my story will provide a record

and legacy that can be referred to by all as we continue to strive for better ways to deal with catastrophic terrorist events that may, and probably
will, occur in the future.

On the afternoon of the 11th, I was watching TV and focused on the
plume of smoke coming from the intense fires. The temperatures were
estimated to be well above 1,000 degrees Fahrenheit, and after the
initial WTC dust settled, the plume rose above the sixteen areas that became known as Ground Zero. If anything was helpful on that day, it was
that the heat released by the intense the fires pushed the smoke upward
about 2,500 ft., and the very steady winds from the northwest moved
the plume over Brooklyn and out to sea. A three-dimensional graphic
of the WTC plume of smoke from the fires is shown in figure 1.4. This
simulation, produced by my colleague Panos Georgopoulos and his
laboratory at EOHSI, clearly shows the plume rising and moving out to
sea. Figure 1.4 is a still from a video that Panos created to help people
understand the potential contacts that might be experienced downwind
from the fires. Based on the plume rise (the height of the smoke), the
people of Brooklyn were spared extremely significant exposures to the
smoke during the twenty-four hours after the attack when the fires were
most intense. If the plume had hugged the surface, the air in Brooklyn
would have been pitch black and barely breathable that entire day. The
area would have looked like the surface-level atmosphere just after a
pyroclastic volcanic explosion.

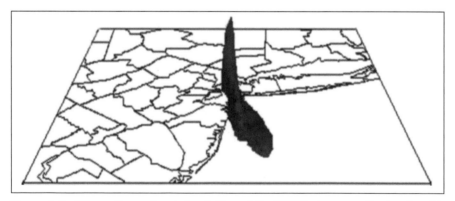

**Figure 1.4. Three-dimensional simulation of the WTC plume on September 11,
2001, at 1:30 PM. Courtesy of EOHSI Computational Chemodynamics Laboratory,
Division of Exposure Science, P. G. Georgopoulos and P. J. Lioy.**

By the morning of the 12th, I began to think seriously about the nature of WTC dust that had fallen on the streets and in buildings, and had deposited on the bodies and clothes of many who were caught in the dust cloud during the collapse of each tower and the immediate aftermath. As I stated in an interview with an archivist from the NJ Historical Society (Louise Stanton, September 27, 2002), my initial thought was that there were going to be occupational and environmental issues. At that time, I was just speculating. I needed to find some way to get downtown to assess the situation. Unknown to me at the time, my colleagues at the NYU Institute of Environmental Medicine had already gathered a number of small dust samples around Ground Zero, and eventually these would become extremely valuable in comparisons with some of the results we obtained from our WTC dust samples and studies conducted or directed by NYU.

THE NATURE OF PARTICULATE MATTER

At this point, we must place the WTC dust, and the closely associated concept of particles that could become airborne and contained in the WTC dust, into a very basic scientific framework. This will help you interpret the information presented later on the characteristics of the WTC dust and the consequences of inhaling it. I will first describe the types of particles that could have been inhaled by victims, rescue workers, cleanup workers, and the public during the aftermath. Then I will briefly describe where airborne particles are deposited in the respiratory system after a person inhales them.

The most important concept to understand is that particles include a very large range of diameters that extend from the size of a raindrop or sand pebble down to sizes that are just a cluster of molecules, detectable only under the most powerful of microscopes. Thus, particle sizes range from millimeters in diameter down to nanometers in diameter. That is six orders of magnitude or a million-fold difference in size from the largest to the smallest particles.

To make things even more interesting, particles can be composed of different substances and can have a variety of materials adsorbed (stuck) on the surface or absorbed inside. On top of that, particles of different sizes behave differently. The larger particles appear to behave like small rocks, falling to the ground quickly, and the smaller particles can behave like large gas molecules: they are capable of staying in the air for long periods of time. Particles can be formed by physical processes and chemical reactions, and they are found in the solid or liquid phase of matter. So, particles have very interesting chemical and physical properties that affect the way they behave, and have been an important area of my research since the 1970s.

For the purposes of this book, I have separated the sizes of particles released by the towers' collapse and subsequent fires into three size ranges that have different characteristic behaviors. A simplified view of the nature of these materials can be described as follows.

First there are *fine particles*; these are in the size range that starts at around 0.1 μm in diameter up to 2.5 μm in diameter. In other words, this means sizes between 0.1 of a millionth of a meter in diameter to 2.5 millionths of a meter in diameter. These types of particles enter the respiratory system through the nose or the mouth and can reach the many circular alveoli sacs that populate what is commonly called the gas exchange region of the lung. Thus, these particles can travel deep into the lung. Figure 1.5 illustrates the main features of the lung and where one would expect to have particles deposited after inhalation through the nose and the mouth. About 20 percent of the fine particles will deposit in the lung if the particles are breathed in through either the nose or the mouth. These types of particles are known to be produced by combustion and industrial processes, and direct formation from gases that are either produced by human activity or nature. The most common example of the latter is the inhalation of smoke released by a forest fire.

Second, we have *inhalable particles*. These can be described by a diameter that starts at around 2.5 μm and extends to about 10 μm. Such particles will primarily deposit above the gas exchange region, within the bronchial airways and the bronchioles (also shown in figure 1.5), and are associated with the thoracic region of the lung. About 25 percent or 10 percent of these particles will deposit in the lung after they are breathed in through the mouth or the nose, respectively. These will preferentially be captured at the points where the bronchial tree branches, including about twenty generations of branching. Also called *coarse particles*, this type is formed primarily by mechanical processes, both natural and hu-

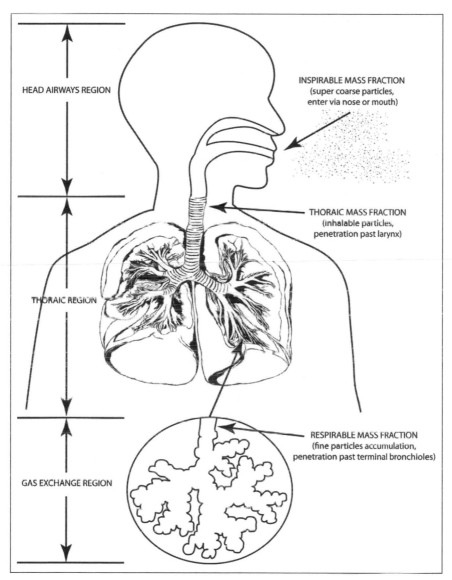

Figure 1.5. The human lung from the point of view of particle size. Courtesy of the EOHSI Division of Exposure Science, P. G. Georgopoulos and P. J. Lioy, and adapted from many sources.

man. Typical examples include dust storms, roadway dust resuspension, and metal grinding.

Third, we have *inspirable particles*. They include the aforementioned types, but most can be described as all the types of particles that are greater than 10 μm in diameter. These larger particles preferentially deposit, up to 100 percent, in the nasal-pharynx (nose and windpipe) regions of the respiratory system. However, small amounts can be deposited in the bronchial region of the lung, and a very tiny amount may even reach the gas exchange region of the lung. To provide a specific classification for the particles greater than 10 μm in diameter, I call them "supercoarse" particles since they are still primarily formed and emitted by mechanical processes, and they are very large compared to what we typically find in the ambient air. The supercoarse particles in ambient air are not regulated by any federal agency at this time, which becomes an important point as we move forward in this story and discuss significant human exposures to the WTC Dust. The supercoarse particles are the ones that you and I have seen every day near a city street, in our homes, and in the countryside; they include street dust and surface dirt.

In addition to the separation of particles by size, particles have another characteristic that is also important for understanding the implications of the WTC dust: shape. The simplest way to describe a particle is as a sphere, which is nice for completing mathematical calculations in textbooks on aerosols, but in reality, particles can take many forms.

For the above, you can tell that particles can come in all kinds of shapes as well as sizes. They can include spheres (hollow and filled), droplets, quasi-spheres with jagged surfaces, and long cylinders. The latter are commonly called *fibers*, thin or thick; asbestos is an example. All of these particles will behave differently when in the air, and the concept of *aerodynamic particle size* (the apparent size at which a particle behaves in the air) will be discussed throughout the book.

The health effects caused by breathing various types of particles can extend from asthma through lung cancer, with cardiac effects and many more respiratory effects in between. Volumes have been written on these topics. The bibliography includes some references for the reader interested in the health effects caused by particles.

THE WEEK AFTER THE ATTACK

With these definitions as a guide, let's return to the story. Two specific events happened on September 12, 2001, that positioned my colleagues and me to be part of the WTC dust issue for over five years. First, just after arriving at EOHSI on the morning of September 12, I was trying to figure out a way to get to Ground Zero to begin the process of understanding and data gathering. Then I received a phone call from Dr. Chris Portier, who was director of the National Toxicology Program at the National Institute of Health. He said, "We're all looking at this and thinking of the event in terms of toxicology. Is there a way to mobilize you folks to go and get samples of the material, because people are going to continue to breathe it?" I told Chris that I had started to think along similar lines, from the point of view of exposure to the local people, and would start to get a sampling approach together.

The second event was receiving a phone call from my colleague, Dr. Mark Robson, and another opportunity to visit Ground Zero began to unfold. He said the School of Public Health of UMDNJ had received a call from the Port Authority of New York and New Jersey, which owned the Twin Towers as well as five other building comprising the World Trade Center, asking for help in identifying the environmental and occupational health issues in the area. It wanted a team to come as soon as possible. We discussed who should become involved with the Port Authority request. A great quality of our institute is the wide range of disciplines that it houses. So, it was not hard to select a team with experience in toxicology, exposure science, public health, and environmental and occupational medicine. The next question was, Who would volunteer to go to Ground Zero?

The team Mark put together included Drs. Michael Gallo, Howard Kipen, Michael Gochfeld, as well as Mark and me. Eventually the group was named the E Team by the *UMDNJ Magazine*. Each member had a particular area of expertise needed for the evaluation, and all could be counted on to give firm recommendations. Michael Gallo trained as a toxicologist with nationally recognized knowledge of chemical carcinogens and other toxic agents. Michael Gochfeld was trained in environmental and occupational medicine and has an extensive background in toxicology, neurobehavioral development, and heavy metals. (Mike also had been deployed overseas in Southeast Asia during the Vietnam War.) Howard Kipen was added as an environmental and occupational medicine expert, with special emphasis on sick building syndrome, occupational pulmonary and hematologic disease, and the effects of exposure to carcinogens. Mark Robson, our leader, is trained in public health and is an expert in the human exposure and health effects of pesticides and rodent control. I rounded out the group because of my work in human exposure science to physical and chemical environmental contaminants, hazardous wastes, and aerosol science.

Mark set about making all the arrangements for our team's visit to Ground Zero. That was a major plus since he is a superb organizer, and facilitator. Personally, I was glad he was the lead since that allowed me to focus on the evolving occupational and environmental exposure issues. With all the rescue efforts already underway, we were going to be another set of volunteers at the site. We believed we would look at the situation with different eyes. In the end, I thought we brought an important perspective by framing what turned out to be a number of serious scientific issues. The ultimate question was, Who would listen?

Mike Gallo also called me on September 12, stating that Tom Sinks from the Centers for Disease Control and Prevention (CDC) had called and was wondering about sampling that was being planned in Lower Manhattan. I called Tom, and we discussed some of the issues and my contacts with the National Institute of Environmental Health Sciences (NIEHS). He was clearly concerned about the lack of coordination. Our discussions did not lead to follow-up with his group, but they started to highlight the lack of focus that was evolving at Ground Zero. He expressed reservations that were similar to mine: we were not sure who was in the lead. Was it EPA or the city of New York or the Federal Emergency Management Administration (FEMA)?

In many ways, confusion at Ground Zero was inevitable. There were rescue efforts ongoing with no idea about the number and condition of potential survivors. We were now at war, but with whom? As a country, we just were not prepared to deal with the environmental and occupational health issues as well as public health issues that can occur after a terrorist attack. For example, before 9/11, I had never attended sessions related to homeland security at any technical meetings. In fact, there were very few sessions on the topic; such sessions generally covered military applications.

On September 12, 2001, my long-term WTC dust- and smoke-related activity began: answering questions and providing explanations to the press. During the morning, I started to receive a number of phone calls about the pungent smell in the air. The first thing I did was ask Panos Georgopoulos, who had started tracking the plume on 9/11, whether the winds had changed. His answer: yes, the WTC plume was making its way across New Jersey. The press started to call looking for an explanation, and then Jeanie called and said the air was "foul" at home and asked me, "Where is it coming from?" I said that she had her first whiff of the smoke from the WTC fires and all that was being turned to ashes at the site of Ground Zero. The calls increased, and my thoughts raced about what would happen if the plume stayed in New Jersey. The short burst made by the plume into Union County caused school officials to restrict outdoor activities, a reasonable decision at the time since we were not sure of the content of the gases and particles in the plume released from the fire. Fortunately, it did not linger over New Jersey, and, for the most part, the worst and the thickest smoke released by the intense fires was dispersed out to sea. I had an interesting call that day from a surfer in Manaloking, New Jersey, about seventy miles south of NYC. She smelled an awful odor; I told her that it was the WTC plume. That person was a former graduate student, who was trying, in her own way, to sort through the horror. By the end of the day, the changes in wind direction provided the entire metropolitan area a chance to smell the plume. The movement of the plume that day was also captured in Panos's 3-D simulation of the WTC plume (still shot in figure 1.4).

On September 12 and 13, we began to get some information on the local response, Tom Sinks of the CDC continued to try to find out what was going on, and sampling plans were under development. However, Tom did indicate that FEMA was the lead. I had some concerns since FEMA hardly, if ever, deals with dust; it usually deals with mud from a flood. Tom indicated that NYC should be in the best position to determine current and future needs for sampling during the aftermath. There was confusion, however, which delayed the discussion or implementation of any overall coherent plan for settled dust sample collection and analysis. In the end, there was none. The EPA was heavily involved with the aftermath at the regional level, and much of their work has been summarized by Matthew Lorber (see the bibliography). The EPA Region II offices were just to the northeast of the WTC Towers (290 Broadway), and I know that many people there saw it all happen. The regional offices were also beginning to get offers of assistance from EPA's National Exposure Research Laboratory in North Carolina.

Tom had been in conversation with Chris Portier, and Chris had indicated that the local universities had analytical and sampling capacities. Tom said that the best plan was for the universities to consider sampling since there were few individuals available from NIOSH or EPA. I agreed with his logic. EPA emergency responders had enough on their hands, and they were taking some spot samples. OSHA and some folks from EPA, for whatever reason, spent much of their time collecting asbestos samples (see figure 1.3), and the experience and skills of the university folks were still on the sidelines. We and NYU were ready to go, and we each found a way to start collecting settled materials.

Some initial air quality tests in Brooklyn and Manhattan did not detect high levels of organic compounds, which would include combustion products from the fires, or lead or asbestos in the outdoor "ambient" air. OSHA also made the unbelievably naive comment to the press that indoor air quality will meet standards for asbestos. There wasn't any data! I was floored by this statement on that day; no one had done any analysis of what was in the indoor dust. In addition, there was still little to no data available about the outdoor dust. Furthermore, what was fueling the fires? Was it jet fuel or other fuels? It was clear to me, however, that we needed to get downtown and collect samples.

The comments from OSHA, the city of New York, and the EPA did lead to a flurry of articles about asbestos levels. Unfortunately, and to the detriment of the public and workers, this coverage focused attention away from a number of other more immediate air pollution issues at Ground Zero and the immediate surroundings. Asbestos is a highly toxic substance, but the health effects from inhaling the fibers develop over a long period of exposure. People were losing sight of the potential short-term effects of personal exposures among volunteers, city uniformed officers, and the public caused by inhaling the WTC dust and smoke. Asbestos never became a *lower* priority—although the scientific issue outdoors eventually became focused on emissions from the pile of debris and rubble, and indoor dust. Due to the asbestos issue in general, no one was able to put the WTC dust in a proper perspective for the general public during the first weeks. I believe the overemphasis on asbestos led to much of the eventual distrust that built up within the community. As Mike Gochfeld said, "They [the EPA and the city] focused on asbestos and confused a working-lifetime exposure with a short-term exposure" (Mike Gochfeld, pers. comm.).

Except for comments made by Mayor Giuliani and others on the lack of weapons of mass destruction (WMDs), the advisories that came out initially were primarily focused on what was available: the asbestos results. As time went on, when it did not appear that high asbestos levels were present in many samples, there was a false sense that the environmental issues outdoors and indoors were straightforward. However, what was forgotten by most, and which is hard to believe, was the total mass of WTC dust with an undetermined composition that was deposited in and around many buildings. Specifically, we did not know the loading of toxic chemicals on surfaces or in surfaces, nor the percent per unit mass of material deposited in the buildings. To put it simply, we did not know how much WTC dust was on the floors, surfaces, and furnishings in residences and commercial buildings. Confusion continued, as the city of New York tried to keep the lead. The city had to relinquish some control in order for EPA to be called in to complete a number of actions.

Many commentators and others have written about who said what when and why. I will try to limit my comments to those based upon my observations and the facts as I know them. However, for further reading, there is the Government Accounting Office (GAO) report published in 2003, referenced in the bibliography, which outlines many issues. In my opinion the report lost some points by not taking full recognition that as a country we were ill prepared to deal with a situation like the collapse of the WTC.

One important point made by a 2003 GAO report is that in the beginning, the lack of effective communication and discussion was very significant among all of the volunteers, emergency responders, and agencies. Communication of messages got better with time. By then, however, the unprecedented nature of the chaos caused by 9/11, the conflicting messages, and the lack of patience of the public and press could not overcome the early problems. I was just happy that I did not have to deal with the daily onslaught of questions throughout the days that followed going downtown. My excursions with the press were much less frequent, and during the first two years post-9/11 they were more informational in nature since I was not part of any regulatory agency.

PREPARATIONS FOR THE TRIPS TO SOUTHERN MANHATTAN

Friday the 14th was the day when the arrangements began to jell for my trips downtown. That morning I received a fax of the letter written by Chris Portier that provided me access to the area south of Canal Street and that would allow us to obtain samples of WTC dust. So, Dr. Weisel, my colleague at EOHSI, and I put together a dust sampling kit to bring to Manhattan. We decided to collect WTC dust samples on Sunday.

To put dust collection into perspective, it is one of the simpler tools that we use effectively in exposure science to obtain information on potential or actual contact of people with toxic or irritating materials that can cause human exposure by breathing or ingestion, with potential health effects. The collection and measurement of the materials in dust is effective in determining the levels of material that are in or on surfaces. Usually, the deposited material can be easily picked up by a person's hand (contact) and then either absorbed through the skin or adsorbed (stuck) on the skin. The material on the surface of the skin is also available for ingestion into the body. Thus, when dust is contaminated with other materials, it can be an important source of toxicant exposure if you touch surfaces with the dust on them and then put your hands in your mouth. Dust can also be inhaled when it is resuspended in the air by motion. Outdoor resuspension is usually caused by the wind or, in urban areas, by the movement of cars and trucks. There was the possibility that settled WTC dust would be resuspended into the outdoor air and breathed into the lungs. This did happen both on the Ground Zero pile and in uncleaned areas immediately around Ground Zero.

MEASUREMENT TOOLS TO EXAMINE EXPOSURE

The dust collection technique generally involves putting a template, usually a square or rectangle, on a hard surface (e.g., floor, table, or window sill), and sampling all the dust inside with a weighed filter or other type of sterile fabric. For some surfaces, this may take two or three repetitions to remove all of the material. Sometimes technicians use a vacuum cleaner modified with a collection filter to sample the surface. The latter is very effective for collecting dust from carpets or plush furniture (e.g., a sofa), which have dust particles present below the surface.

In contrast to dust collection methods, other instrumentation that we use in the field to measure exposure are more sophisticated, including personal monitors (i.e., monitors worn by a person). Unfortunately, these were not available or could not be practically applied during the immediate aftermath of 9/11. Some of these sampling devices are attached to a person's clothing, with a sampling pump that is also attached to actively sample and monitor the air. Or, the sample can be collected by using the principles of diffusion of molecules. If time is available before something happens, such as an accident, or other type of dust release, dust can be captured passively (without a pump) as it deposits on a collection plate that is not affected by rain or snow. This can be as simple as an open plate with no contaminants on it with a nonelectrostatic surface, like the material used to pack computer parts.

Sometimes the personal sampler is a direct reading device and records the toxicant levels immediately on a minicomputer or thumb drive for eventual analysis. Other times a device will collect a sample of the chemical, radioactive, or biological agents of concern over a period of time on a filter or in an absorbent material. Most of these samples will need to be sent back to the laboratory for analysis. In addition to personal monitors, we can also collect samples with active or passive monitors in a stationary location that is representative of the air we breathe (near a street), or the water we drink (tap water from the home), or the food we eat (e.g., a duplicate plate of the food eaten at dinner). Each is then analyzed for the contaminant or pollutant of concern.

For the air we breathe, these monitors must be portable and not require an electrical connection, something that was in short supply near Ground Zero.

For other types of exposure studies, the urine, blood, breath, or hair of a person can be collected, and then the levels of toxicants that a person may have contacted over a short or long period of time can be measured in the samples. Normally, this approach is used in medicine for routine analysis of bodily functions. For example, urine analyses are performed to indicate kidney and liver functions. In all cases of exposure science, the goal is to examine a person's contact and exposure to a toxicant (e.g., lead or arsenic) that entered the body by a single or multiple routes. Included is eating, drinking, inhaling, and dermal (skin) absorption of the chemical, physical, or biological material. For example, most children living in cities or older suburbs may undergo a routine test of their blood for the presence of lead before the age of two. The technique is called *biomonitoring*.

During the WTC aftermath, Dr. David Prezant of the NYC Fire Department measured the levels of toxicants in the urine and blood of firefighters. This was very useful for examining their exposure since the fire department also took samples from firefighters not at Ground Zero. These are called *controls*. Therefore, they had baseline levels on toxicant loadings in the body of FDNY personnel. It is my hope that eventually we routinely measure toxicant levels in the urine or blood of all people who take annual physicals to catch environmental insults early. However, this tool is years away from reality. To become a meaningful component of "personalized medicine," each test must be able to be interpreted meaningfully for each patient.

During the 9/11 aftermath, we did not collect airborne or resuspended dust samples because of the delays mentioned previously and the lack of the correct types of portable air monitors. The best that could be done was to collect the settled dust. This dust would represent material people contacted during and after the collapse, and material that was or could be resuspended into the air by emergency vehicles or the wind.

2

COLLECTING DUST AT GROUND ZERO

The arrangements for the Port Authority authorized visit of the E-Team were formalized by Mark Robson by midday on September 14. Mark called to say we would be leaving early Monday morning (the 17th) to visit Ground Zero. That seemed to be a reasonable plan.

Within a week after Chris Portier's letter was sent to me, his direct role in the WTC began to diminish. The National Institute of Environmental Health Sciences response to the WTC moved into the director's office, Dr. Kenneth Olden. However, I am deeply indebted to Chris for having the foresight to know, on the federal level, that it was very important to collect and analyze the WTC dust that was on all surfaces from the beginning of the aftermath. Furthermore, he also noted that the local universities were an asset to the response. As noted in chapter 1, colleagues at NYU also saw the wisdom of collecting dust samples and did so, even before us! Lung Chi provided me with a unique picture taken by NYU colleagues on the 13th (see figure 2.1). It is a visual image of the postcollapse "blizzard" of dust that had settled in Lower Manhattan. It is a chilling picture since before the collapse of each tower, no one would have ever considered such a possibility, except after a Mount St. Helens–type volcanic eruption. Family friends Bob and Lori Saunders went into the city on the 12th and worked their way downtown. They saw inches of WTC dust on cars starting just above Canal Street at 189

Figure 2.1. The post "blizzard of dust" deposited on surfaces in southern Manhattan and available for resuspension into the air. From a series of photos taken by Dr. Mitchell Cohen and Ms. Colette Prophete, NYU Institute of Medicine. Reprinted with permission.

Centre St., a point never fully appreciated by agencies. Bob and Lori did not touch the dust since it was not a familiar material, and they were uncertain about what was in it. EPA and others collected dust during the initial days but completed either a less extensive or a more specific set of analyses.

Early Sunday morning on the 16th of September, I met Cliff Weisel at his home and we drove into Manhattan. It was very difficult to get around, but we managed to park south of Canal Street, which was defined during the aftermath as an upper boundary for entry into Lower Manhattan. Having the NIEHS letter in hand, we were given permission from an NYPD officer to enter a controlled area around Market Street and then collected two large bulk samples of WTC dust, over one pound each. I did not realize it at the time, but these were the only two known samples taken between the Brooklyn Bridge and the Manhattan Bridge near the East River. To me they represented a burst of dust that "hopscotched" into the area because we did not see WTC dust everywhere. Some of the dust deposited in the area may have been washed away by the Friday rain, and we found similar, but smaller, deposits in other locations. For example, there was WTC dust on the hoods of cars, which provided support for my original conclusion about the burst. A picture of one car hood we sampled is provided as figure 2.2a. Figure 2.2b is a close-up picture of the bulk appearance of the WTC dust—a major cause of health concerns after the WTC collapse. I have donated a WTC dust sample to the New Jersey Museum, which is now in its WTC archives. It can be accessed for viewing by the general public.

Both Dr. Weisel and I were very much aware of the growing concerns mentioned in newspaper articles about WTC dust and the WTC smoke plume from the continuing fires. Therefore, after collecting two large WTC dust samples, we decided to drive across the Manhattan Bridge into Brooklyn to look for and collect samples of WTC dust. We made an extensive trip along the streets near the East River, but we found nothing. Maybe the rain had washed the outdoor settled dust and smoke away.

During the months ahead, we were still trying to obtain WTC dust samples from residents in Brooklyn. In fact, eventually I was willing to analyze anything that people thought was WTC dust. However, over the course of a year, and even after a plea in *Newsday's* August 23, 2002,

Figure 2.2a. Car surface of WTC dust sampled by Drs. Weisel and Lioy on September 16, 2001. From a collection of pictures taken by Paul J. Lioy.

Figure 2.2b. The actual WTC dust sample collected by Lioy and Weisel. Electron micrograph taken for MVA Project 4808, MVA Scientific Consultants, GA. Lioy et al. *EHP* 110 (2002): 703–714.

edition, during an interview with Laurie Garrett, I received nothing that resembled WTC dust from Brooklyn. Eventually, we did receive some very small samples from residents of Manhattan, and some of those did have WTC dust.

So, we went home that Sunday, satisfied that we had collected samples. Now the task was what to do with them. Years later, I did say to Cliff that I had made one mistake that day. We had the opportunity to move south and west of the area toward Ground Zero, but instead we decided to get in the car and go across the bridge. That was almost a fatal scientific mistake since we did not have dust from immediately around Ground Zero. Fortunately, that mistake was rectified during the Port Authority trip that occurred the next day.

On September 16 and 17, the United States Geological Survey (USGS) collected samples at radial distances from Ground Zero, all below Canal Street. But its efforts were uncoordinated with us. In addition, when we lost the opportunity to collect samples along the path to

Ground Zero (the sixteen-acre site of the former towers) from Market Street, we lost the ability to eliminate controversies during the next four years about the distance the dust moved from Ground Zero into the local neighborhoods north and east of the site. Furthermore, we did not get a chance to share samples and complete complementary analyses on the same samples. The lack of such information did not help the community's perception of the poor design of what was actually a reasonable indoor cleanup program eventually developed by EPA. Personally, I could never understand why I passed up that opportunity to walk to Ground Zero on the 16th, and I will never hesitate again in future studies of public health issues.

As a note, the 16th and 17th were important days in the broader scope of sampling during the aftermath. During this period, the EPA finally received approval to start sampling in earnest at well-situated sampling sites around Ground Zero for materials other than asbestos. It was a pity that we did not know all this was happening prior to going downtown since collaboration could have led to more samples and maybe even a more general sampling design. The 17th was also the day the workers and commuters were allowed to return to Wall Street, although not by traditional routes such as the Holland Tunnel or the WTC Path trains from New Jersey.

We will talk about the USGS work later, but you should know that its WTC dust results were available to FEMA, NYC, and the EPA by September 27, a full two weeks before ours were beginning to be discussed, but its report was not released by agencies to the public until late November. In hindsight, this was a mistake on the part of FEMA and the city, as confidence in the state and federal agencies continued to erode in downtown New York City.

I could not rest on Sunday's successes or lost opportunities. I had to prepare myself for the next day, when the E-Team was scheduled to go to Ground Zero. The media continued to show the situation downtown during this stage of the aftermath, and the intense reporting continued to give me reasonable insights about what to look for at Ground Zero the next day. The E-Team left for New York City at about 5:30 AM and arrived just after sunrise. It was an eerie morning—the sun was out, no clouds, sort of like the morning of the 11th. If you did not know where you were heading, you would have thought it was a very nice late summer day.

It took a while to make our way downtown; again, many streets we took were closed, but our driver was excellent. He dropped us off at Canal Street, where we waited for a representative from the Port Authority. I was impatient, but that is my nature, and I was somewhat focused on getting a WTC dust sample from a location immediately around Ground Zero. We finally started downtown and got some gear at Manhattan Community College located on the west side of Lower Manhattan. The biggest surprise was all of the food and materials that had been acquired in such a short period of time and were readily available in the staging area to all of the volunteers. Yes, there were respirators, bins of respirators. We donned ours; others did not. This was a theme that was to be played out throughout the days after the collapse of the Twin Towers. Mike Gochfeld's overall comment was that "at our visit we were virtually the only ones wearing respiratory protection. We saw lots of people with respirators dangling around their necks. We saw a large 'warehouse' with people able to pick and choose their own respirators without any guidance. Where was industrial hygiene?" (Mike Gochfeld, pers. comm.). Even in our situation as invited reviewers, there were no arrangements made for trying to have one of the occupational health agencies fit-test the respirators to our faces. Fit-testing reduces the possibility of allowing toxic materials to seep between the mask and your face. A proper fit prevents particles and gases from being breathed in by the person wearing the respirator. So, the devices were somewhat compromised, but at least we had them.

The respirator and headgear also have symbolic meaning for me. In 2008, during a conversation with the EPA's Dan Vallero, who was one of the leads for the National Exposure Research Laboratory at Ground Zero, we were discussing the WTC experience. He mentioned the fact that he still had the steel-toed shoes he wore at Ground Zero in a plastic bag at his home, and they were located in a special place. He had not touched them since his work there ended. Thinking about my own gear, the headgear is on a shelf in my office and has only been moved twice to be cleaned. The steel-toed shoes I wore are on a shelf above a work bench at home, and these have not been used since my last sampling trip to Ground Zero in late November 2001.

The E-Team was driven the rest of the way to Ground Zero by Port Authority personnel, and we met another group from Mount Sinai

School of Medicine. Just as we were getting ready to put our gear on, the absurdity of some people or denial of the event came into full focus with an ironic "New York moment": a man dressed in only shorts and a T-shirt ran past us completing his daily jog, and he was on his way downtown! Here we were walking with respirators and hard hats on, and this person was running straight toward Ground Zero. I wonder if he ever complained about respiratory effects or post-traumatic stress. Did he read the newspapers? I wondered at the time, did he really care about what happened the week before? I wonder if he ever figured out that this was war! However, this moment also provided a good object lesson for me when I had to deal with the press and public in the future about the health effects associated with WTC dust and smoke released on and after 9/11: you must expect the unexpected in human behavior after a traumatic experience.

Recalling his first thoughts as we entered Ground Zero, Howard Kipen said, "As we approached the site I was in awe of the enormity of the physical change in the environment, of the accumulations of dust, of the strange odor from the burning pile, and of the enormity of the complex response with activity all around, and yet a calm sense of people just doing their jobs" (Howard Kipen, pers. comm.). The rest of that day included a "hands-on" systematic review of the entire scene at Ground Zero. The E-Team started on the west side of the sixteen-acre plot of land that was now filled with twisted metal and debris, and we moved southeast and finally to the north. Each part of Ground Zero presented another feature of the rescue efforts during the aftermath and challenges for the cleanup and restoration. Figures 2.3a–d are representative of the time we were moving around Ground Zero; each angle gives a different view of what the rescue workers and demolition crews were dealing with that day. There was no true order, but there was no chaos. It seemed like individual agencies and contractors took charge of specific locations. Since the Port Authority was the owner of the towers, we felt confident in having access to all of the places that were necessary to conduct our review.

I felt many emotions as we walked through the area examining issues of importance. About two-thirds of the way through our review of Ground Zero, I had my opportunity to take my third outdoor WTC dust sample. It was just before we passed the Millennium Hotel, about

Figure 2.3a. Looking east next to the pile, September 17, 2001. From a collection of pictures taken by Paul J. Lioy.

Figure 2.3b. Looking north and east, September 17, 2001. From a collection of pictures taken by Paul J. Lioy.

Figure 2.3c. Looking east, September 17, 2001. From a collection of pictures taken by Paul J. Lioy.

Figure 2.3d. Looking south, September 17, 2001. From a collection of pictures taken by Paul J. Lioy.

a third of the way up Cortlandt Street. This sample was obtained at a location very close to Ground Zero, within about two hundred to three hundred feet. The location had been sheltered completely from the elements, including the rain, for six days. It was still an undisturbed outdoor sample. This point will become important in my discussion of the "alkaline" (basic, like Drano) nature of WTC dust samples.

We reviewed many occupational issues, and it was obvious that only a few people were wearing respirators. I took many pictures, but the classic for respirator use on that day was of three rescue/recovery workers standing in a row (center of figure 2.4). The first worker was wearing his respirator, the second was wearing it around his neck, and the third had it on his belt. Clearly, the many messages about wearing respirators were not being heeded by all of the workers, and this was even before the controversy about the air being dirty versus clean hit a crescendo on September 18. Early on, the EPA, not the city or OSHA, was telling workers at Ground Zero to wear respiratory protection. (This point is made in Juan Gonzalez's book *Fallout*.) As mentioned, I was wearing a respirator, but since it had not been fit-tested, I did get multiple whiffs

Figure 2.4. Three workers (center next to the pile) and respirator use at Ground Zero on September 17, 2001. From a collection of pictures taken by Paul J. Lioy.

of the distinctive odor that Jeanie had smelled on the 12th. I was probably not totally protected from the dust.

With the approval of the Port Authority, we all took many photos in and around Ground Zero. Many were of worker tasks and the general conditions of the surrounding area. However, just to give you an idea of the volatile nature of the situation, I offer the following encounter. As we were preparing to leave the Ground Zero area and our team was walking toward the Manhattan Community College staging area, I was still taking pictures. A number showed how the dust was continuing to stick to windows even after the Friday rain. Then the E-Team turned a corner, and in front of us were members of the military. An officer came up to me and bluntly said, "No pictures," with the warning that I could be detained for twenty-four hours if I did not stop. I did not argue or ask questions; I just put the camera away. It was clear that they were in charge of the block, and I had to obey their rules. So, just one block over from the Port Authority area, I was in another area of authority, a different world. Thinking back, there were probably other areas of "authority," but we may have been steered clear of those by our guide.

As I reflected on the experiences of that day, to my amazement and pride, I realized it was a chaotic scene, but not one of confusion and hysteria. Mike Gochfeld provided the following observation: "My impression at Ground Zero is that there were seemingly a dozen heavy equipment sectors, with cranes swinging everywhere while people were balancing on 45-degree angle beams, high above the pile. I had anticipated that there would be fatal accidents among the rescuers, but it all seemed to orchestrate itself, and in the end there were far fewer events than expected" (Mike Gochfeld, pers. comm). All organizations present were working toward two goals: rescue and then movement toward recovery activities.

The one unfortunate aspect of the scene was the lack of coordination among groups, including environmental and occupational health issues. Based on my areas of expertise, the most significant deficiencies were (1) the lack of mandatory use of the respirators, (2) no well-defined personal monitoring of the workers, and (3) the lack of a coherent plan for environmental and occupational sampling and analysis during the early phases of rescue, the days before September 17. Remember, this last point had been mentioned to me by CDC on the 12th.

We felt that an overall occupational health and safety program needed to be implemented with a single agency in charge. On the day we were there, over half of the workers were not wearing respiratory protection. In the buildings where dust was being swept out, many of the workers were only wearing paper dust masks, if that. The lack of firm and well-defined respiratory protection measures contributed to the health situation that evolved down the road for workers and volunteers that were on or near Ground Zero at the beginning of the aftermath, and through the first few days of rescue efforts. To repeat: This consequence was not due to the lack of respirators at the site. It is a pity that I did not take a picture of the bins of respirators. Who would have known the value of such a shot at that point in time? Occupational health agencies were slow in getting a mandatory respirator fit program started at Ground Zero. In my view, the people in charge were just not taking the overall acute inhalation risks seriously, or they were adhering to nonwartime policies about who is in charge of a rescue scene. A respiratory protection program was finally implemented by OSHA in collaboration with Mount Sinai School of Medicine in October, a month after the collapse.

Here is a quote from the 2003 GAO report (see the bibliography for the full reference) on the respiratory protection issues:

> Our limited work in this area indicated that respirators were generally available but they were not worn for a number of reasons. A significant factor was the desire to save lives without regard to personal safety in the immediate aftermath of the disaster. Other reasons appeared to include the respirators' interference with the ability of emergency workers to communicate, lack of training, lack of enforcement of safety measures at the site and conflicting messages about the air quality at Ground Zero. (U.S. EPA Region II 2003)

I would add that misinterpretation or overinterpretation of messages about air quality at Ground Zero contributed to the lack of respiratory protection being used in the beginning of the aftermath.

Today, however, I still do not understand why the many people working or volunteering at Ground Zero, or reentering their homes, did not take precautionary measures *on their own* to prevent inhaling WTC dust. Simple logic would dictate that if you go into an area of high dust loading, you should have some degree of protection. If you experienced problems, get some advice about what you need to do. No one has answered this question to my satisfaction, and probably no one will. However, the same comment can be made about those who do not listen to, or fail to heed, mandatory evacuation orders during a hurricane.

The risks from physical injury were large at Ground Zero. I had one brush with a serious accident while observing the pit at Ground Zero. While taking pictures of the debris removal and the smoldering fires during the rescue efforts, I was standing over the pit, looking down. I did not see an I-beam being moved by a crane, and it was coming straight for me. I had a "buddy" from the rescue team, and he screamed, "@&#!! Step back, Doc!" I did, but without a megaphone built into my respirator or his it was hard for both of us to communicate. So, we took them off to keep communication going and keep me out of the emergency room. As I was leaving, I took a picture of the beam that had almost made me an item on the evening news.

Obviously, there was a problem in the ability to communicate while wearing respiratory protection. In 2008, communication difficulties continued to exist for non–air pack respirators, the kind we both were wearing at the time. (Air pack respirators require a cylinder of air to be worn on the backs of rescue workers. These are seen in many movies depicting firefighters entering a building.) I am bewildered that we still do not have a selection of non–air pack respirators, those that do not need an air pack and that have a built-in communication system. The air pack respirators have a good communications system. Mike Gochfeld and I made this recommendation soon after the event and published it in a lessons learned article cited in the bibliography. It is a pity that with all of the money spent on buying trucks and gear by the Department of Homeland Security, the investment into developing a better respirator is still a goal! This is a major challenge for the next generation.

Another immediate concern we identified at Ground Zero was operations in confined spaces (such as remnants of basements and PATH tunnels). One example was the concern for presence of containers of Freon and any other compressed gases underground. It is just remarkable that there were few major accidents, a testimony to the professionalism displayed by all of the heroes involved with the rescue efforts. This happens because of experience and training, not only by luck. Finally, we were concerned about decaying food that could cause disease if not cleaned up quickly, but fortunately, this did not become a major problem.

The review by the E-Team did not lead to unanimous conclusions on the issues and what should be done next. For instance, initially some of the team did not agree with me that the resuspendable WTC dust would become an important health problem for workers. However, we all agreed that asbestos should not have been the only major focus, and the lack of WMDs was a blessing. Our report to the Port Authority was well thought out, and we eventually reached a consensus. What happened to the report, I will never know. Mark sent it about two days after the E-Team visit. We never got a response from the Port Authority. This was a major disappointment. An even greater disappointment was the fact that none of our recommendations were ever acted on by the Port Authority, with, of course, the most serious disappointment being the lack of a firm mandate to all workers about wearing respirators. Our recommendation was made, the EPA continued the advisories, but the boxes full of respirators on the 17th told, in the classic words of Paul Harvey, "the rest of the story."

Personally, the visit to Ground Zero on September 17 was the education I needed to prepare for the months and years ahead. It gave me a firm understanding of the reality of the situation and a better grasp of critical issues. Without this experience, my value in future discussions and analyses of the environmental conditions indoors and outdoors would have been diminished considerably. A virtual tour on TV or a virtual laboratory experiment is not a replacement for the real thing.

After returning to New Jersey that day, I immediately did two things: (1) placed the dust samples in the EOHSI cold room, where the temperature was slightly above freezing for eventual processing, and (2) dropped off my rolls of film for development. Each of these two events led to immediate but unforeseen reactions.

The first concerned the pictures I took at Ground Zero. I received a call from Mark later that evening; he said that we were being requested to hold off on developing our photos due to the sensitive nature of the subject matter. However, it was too late for me to comply with that request. I had submitted my two rolls of film for processing as soon as I had returned home. This recommendation, however, was curious. Again it reflected the nature of the times. There were reporters taking pictures everywhere around Ground Zero, and the images were being sent immediately around the globe on TV, posted on the Web, and recycled frequently by the media and those who had access to the Internet. Why the request for a delay? What did we see that no one else saw? Again, who was in charge?

In the end, no person from any organization ever officially requested to see the pictures developed by any member of the team, which was similar to the nature of the response to our written report to the Port Authority. We have published many of the pictures, and we have presented them at forums around the globe. I have also provided some previously unpublished and published photos in this book. I often refer to the photo of the three rescue workers on the 17th (figure 2.4), because it captures the many levels of attention—or lack of attention—individuals were paying to respiratory protection.

Before I get to the second event, on the day after our trip to Ground Zero, September 18, EPA administrator Christine Whitman made the statement in an EPA press release that "[w]e are very encouraged that the results from our monitoring of air quality and drinking water conditions in both New York City and near the Pentagon show that the public in these areas is not being exposed to excessive levels of asbestos or other harmful substances" (U.S. EPA 2001). Further information on the situation can be found in the GAO report from 2003 (see the bibliography). The statement goes on to discuss the plans for air sampling that were to be started, augmenting the recently acquired samples. The EPA statement on the 18th goes on to state a point that has been totally lost: "the highest levels of asbestos have been detected within one-half block of Ground Zero, where rescuers have been provided with appropriate protective equipment" (U.S. EPA 2001).

Finally, the EPA statement makes points about wearing face masks and cleaning debris off clothing separately. It is curious, however, that no one appeared to absorb the whole message. The public was being told it was safe to walk the street, not the pile workers. The failure of many people to communicate information effectively and the failure of others to listen carefully clearly fostered the problems that grew as time went on.

Unfortunately, the EPA statements could not affect the WTC dust exposures that had been experienced during September 11 and 12, which would become the most important days in terms of worker health problems. Finally, the sampling devices were just not available at the time to record the necessary data. In light of the types of data available for measuring ambient conditions, Whitman was correct. This was based on the fact that only seventeen of seven thousand outdoor asbestos samples were above the clearance value for safe reentry into an area with asbestos. Further, based on analyses made by Matt Lorber of EPA, benzene, a known human carcinogen that causes leukemia, had elevated levels near the pit area of Ground Zero on the days of sampling during the first days of the aftermath and in the period that included September 16. High levels of benzene were not found in the community environment. Fortunately, all levels of benzene were reduced rapidly as the intense fires died out. Exceedances of occupational health and safety limits only occurred on three days, and by the time the general public returned, there was "no public health risk" for benzene. Overall, I believe that Whitman's statements were misinterpreted since they were directed at outdoor air quality and not occupational issues on or near the pile, and they stated nothing about the potential indoor issues. Misunderstanding or poor communication plagued the aftermath and damaged the relations among everyone. However, I must add that people were anxious to hear that things were better and would listen accordingly. In the future, we just need to listen better and ask questions when we do not understand.

3

FREEING THE WTC DUST

During the days immediately following our visits to New York City, I attempted to start the analyses of the settled dust samples. This got interesting very quickly after September 18, and even before we separated out any portions of the WTC dust for analysis by various laboratories! As I mentioned, Mike Gallo had started talking with the leadership at NIEHS about developing a coordinated environmental health response to the aftermath of the terrorist attack. The agency wanted to get involved as early as possible to begin to deal with the nature of environmental and occupational health issues. Mike worked directly with Dr. Allen Dearry, assistant to the director, and Dr. Kenneth (Ken) Olden, the director, to begin a process that would lead to an academically driven path forward on the exposures and health effects that might be of concern. Their approach was to use existing NIEHS centers of excellence, especially those in the New York metropolitan area, to gather information and provide analyses on the critical health and exposure issues. After the fits and starts that were experienced by NYU and us in our attempts to work downtown during the first few days, I looked forward to more organization.

Mike, as the director of our NIEHS Center, began the process of coordinating the local efforts with the NIEHS. By that time the director's office of NIEHS had taken over the agency's activities associated with the response.

To my pleasant surprise, the engagement of NIEHS in the aftermath was progressing quickly and becoming the main focus of my attention immediately after our E-Team visit to Ground Zero. However, I was blindsided by an unusual question about the dust samples posed by Mike Gallo and Ken Olden: are the dust samples that we had acquired evidence from a crime scene? That thought had not crossed my mind. The issue was DNA, and whether DNA would be present at forensically useful levels in the dust. It was a reasonable concern since WTC Ground Zero was a crime scene and a war zone, and people had died during the collapse of each building.

RESEARCH THAT SHAPED OUR WTC WORK

Before we move into the topic of DNA, I need to digress for a moment to discuss my background and periodic interest in aerosols and subsequently dust. During the developmental stage of my career, I changed from plasma physics, which is the study of highly energized gases like those one would find in a laser or your flat-screen TV, to environmental science. This was not because of a revelation about the importance of the environment, it was due to the simple fact that the field of physics was drying up in 1970, post–the landing on the moon. When Clean Air and Water Acts were passed by President Nixon, the environment emerged as a new area of opportunity. As I said previously, it was Jeanie's idea to pursue the environmental science program at Rutgers University, which existed for over forty years prior to the above legislation. I worked in a small laboratory on the Rutgers campus that included three other coworkers. One was Jack Glennon who died too young, and another was George Wolff, formerly of General Motors, who has made many contributions to air pollution research.

Because of my background in plasma physics, a logical approach to the field of environmental science was to move from exploring the exotic world of ions and high-energy states to tackling significant problems in air pollution. I gravitated to mixtures known as aerosols. Simply put, these have two components: particles and gases, with the particles suspended in the gases. Examples are the mist created at the beach from a wave that moves inland after the wave gives up its energy on shore. In addition to sea spray, the movement and fate of other aerosols fascinated me. I could create another type of aerosol driving my car down a dirt road. Today, I would call that dirt road a source, but then it was just a source of teenage amusement. I would drive my car at about thirty-five miles per hour or more for at least one to two tenths of a mile. Upon stopping, I would turn my head and look back behind to see a dusty haze. In fact, I could see nothing through the haze for ten to twenty seconds. That haze was an aerosol.

Over the years, my research shifted from mainly studying the outside, ambient aerosol to indoor (e.g., dust) and personal (indoor + outdoor + occupational) aerosols. Thus, after I began to work in the environmental sciences, I went from studying only particles that were present in the air and could be physically observed through an electron microscope, to particles that were as large as soil/dirt, consumer products (starch), or human hair, for example. The entire range of particle sizes and shapes and composition became part of my research, which became important for interpreting the results obtained from the WTC dust samples and for understanding human exposure around Ground Zero.

During the 1990s, dust in general became an important area of research for my division at EOHSI because of major state and national issues. We became involved with three major issues during that time: the presence of lead in homes, the methods used to apply pesticides in homes, and the chromium wastes used as fill material in residential settings found in Hudson County, New Jersey. Little did we know that these experiences would become useful in our response to a terrorist attack.

For lead, the scientific issue we examined was the application of a reliable method of measurement in the dust that could be related to lead accumulated in a young child's blood. We also wondered how such methods for testing lead could be used to illustrate source reduction or elimination. We completed studies in homes that related dust levels and cleaning practices to the levels of lead present in the blood. We found that with a good cleaning program, the blood lead of children who spent a lot of time in a home could be reduced by an average of 17 percent, and some homemakers were so well engaged that in one case the blood level was reduced by over 30 percent—a fantastic success! That study involved multiple graduate students and collaboration with my colleague Dr. George Rhoads of the UMDNJ School of Public Health.

In the pesticide studies, we broke new ground for the control of children's exposure to pesticides. Later it was called the "plush toy" study, and my main scientific collaborator was Mark Robson. The research was completed primarily by my graduate student, Somia Guranathan. The study at first was somewhat baffling. We sprayed the pesticide in a routine manner on a surface and placed an unused toy in the same *unoccupied* residence. After one application of the pesticide chlorpyrifos to the room, over time the pesticide levels got higher in the toy; however, the toy was never sprayed! Normally I would expect the pesticide levels to do the exact opposite—decrease—since after the single spraying the pesticide would be removed from the room air, and the rest would eventually degrade or be removed by tracking or vacuum cleaning. Thus, I can say that the observation had not been reported before and was not simple to explain. In fact, I put the results away for a while, and Somia ran some more field tests in the unoccupied residence. Her additional experiments provided a much better picture of what was really happening. The pesticide was not just part of the dust on the surface after spraying. It was a semivolatile material! That meant that after spraying, the pesticide would evaporate and then recondense on another surface. However, plush and squeezable toys had a unique feature: each was filled with polyfoam. During the effort to develop stationary air samplers for indoor and outdoor measurements, polyfoam was found to be a great absorbent or sponge for pesticides. This was an important point for me to remember, as you will see. So, in a real home a pesticide spraying would be done by a professional or someone in the house, and as time went on the toy would act like a sponge and passively fill with the pesticide. This sudden increase of pesticides in and on the toy could occur over a period of days after just one spraying! Interesting physical chemistry, but not good news for young children who mouthed their toys. The results from the study would never allow the thinking man or women to ever again consider pesticides as just residues that hang around on surfaces waiting to be picked up by a hand, toy, or food. Semivolatile pesticides, just like many other materials, are dynamic and can play hopscotch from material and location to material and location. In the paper, we called it an indoor application of a scientific phenomenon called the "grasshopper effect," which explains the redistribution of persistent organic compounds from one place to another. After we published the study in *Environmental Health Perspectives*, I was interviewed by popular women's and family magazines, and our work eventually led to the elimination of what was called professional broadcast pesticide spraying in homes. This has helped to reduce exposure and lessen the contact of young kids with pesticides. The issues of semivolatility would also become part of the WTC dust analyses.

The third issue, chromium, actually provided much of the work that would enhance my understanding about how dust moves and accumulates indoors. This experience was also necessary for dealing with the phases of the WTC aftermath. It also led to the development of a field sampler called the LWW surface wipe sampler (LWW for Lioy, Wainman, and Weisel), which was also used in our lead studies. In this instance, we were studying the exposure of residents in Jersey City, New Jersey, to the element chromium, and we were sampling residences that were located next to, or on, known sites with chromium contamination.

The history of the problem is interesting. From 1905 through 1975, Hudson County, New Jersey, including Jersey City, was a center for chromate production and manufacturing, including chrome-plated bumpers. The facilities generated over two million tons of waste until about 1960. This waste was used as "apparent" clean fill on residential, commercial, and industrial locations. During the late 1980s, people began to see "yellowish" chromium crystals in walls of homes, schools, and other buildings, and there were yellow/green blooms of chromium in soils outdoors. The chromium had a valance state of +6, which is the carcinogenic level form of the element. (Valence electrons are in the last shell or energy level of an atom, a +6 is missing 6 electrons.) By contrast, chromium with a valance state of +3 is an essential element for life. Our goal was to study the pathways by which people were exposed to chromium in locations very close to where they lived.

We sampled the dust present in many homes and found that locations near the land contaminated with chromium waste had higher indoor levels of chromium. These high levels were due to resuspended dust, dust blown off the surface of the waste sites with high chromium, or the tracking of the chromium indoors by residents, friends, and pets. In some cases, we found the classic greenish-yellow glaze of hexavalent chromium on the basement walls. This would indicate that the very mobile hexavalent chromium in the fill had moved through the soil and traveled through the cement walls into the building.

Each of these three research activities helped develop the expertise of the EOHSI investigators on dust issues. I even completed a project that evaluated the collection efficiency of typical household vacuum cleaners for large and small particles that could be deposited in a carpet. Eventually, Jeanie suggested that I buy vanity license plates with the name "Dr. Dust" on them. I have not done so yet, but the thought has crossed my mind.

This type of research placed us in a position to define the forensic aspects of dust within the field of exposure science. Some of these opportunities also introduced me to Dr. Jim Millette of MVA, Georgia, who runs a superb microscopy laboratory. He did a lot of work after 9/11 on the WTC dust and became an important member of our team of scientists who ended up analyzing our WTC dust samples.

DNA AND DUST

As mentioned above, I was beginning to get concerned that we would not be able to analyze the three WTC dust samples because of the DNA issue. Personally, I thought the DNA was not an issue for the samples, because with the ten million tons of building material that returned to dust, and even with the then-projected six thousand or more people who lost their lives in the collapse (eventually reduced to just under 2,900, with a number of bodies intact), the likelihood of finding material associated with human remains in a sample was less than one in a quadrillion (1,000,000,000,000,000)—which would indicate a very low expectation or probability of finding DNA in any one sample. Furthermore, over the twenty-five years of WTC occupancy, there would be hair follicles from thousands of occupants and even more from visitors to the WTC Towers present in the rugs that were crushed during the collapse. However, there will always be an emotional component to what was in the WTC dust, which I could totally understand and empathize with then and now. Even as late as 2005, individuals were still worried about the presence of detectable human remains in dust. I spent a half hour on the phone while at a meeting in Arizona explaining the above to a member of the southern Manhattan community. I believe that DNA was one of the major reasons why the WTC dust remained a fascination to many people. For example, Anthony DePalma wrote a front page *New York Times* Metro-Section article entitled "What Happened to That Cloud of Dust?" on November 2, 2005. On September 5, 2006, pictures of the dust and basic description of the components of the WTC dust were part of another *New York Times* article. For the latter, the focus was on the health effects being studied among the workers and volunteers.

I relayed the results of my estimation for the presence of DNA in a sample to NIEHS, which seemed to place the issue into the appropriate perspective. With the matter resolved, I thought that we could begin to focus on the analysis of the samples. I was wrong! A member of our faculty started to worry about the need to get Institutional Review Board approval for use of human subjects or human samples to analyze the dust. His reasoning was again based on the potential for the presence of human remains in the dust. This claim appeared to be out of bounds; however, Mike Gallo and I had to again go back to Ken Olden for a determination, which meant another delay before we could analyze the WTC dust samples.

By about September 26, we had our answer in an e-mail from NIEHS. The common rule on human subjects states, "Human Subject means a living individual about whom an investigator conducting research obtains: (1) data through intervention or interaction with the individual or identifiable private information . . . private information must be individually identifiable" (NIEHS, pers. comm.). According to the definition, deceased individuals whose remains "could" be in the dust did not qualify as human subjects. Therefore, the institutional ethics review and approval was not required. Thus, we could finally plan to have the samples analyzed for constituents in the dust. However, we had acquired a much greater sensitivity toward the broader interpretations that people could ascribe to the WTC dust. We had also lost precious time that could have reduced confusion about the significance of the WTC dust exposures.

FIRST THOUGHTS ABOUT ANALYSIS

With the question about the WTC dust finally resolved, I began to call friends and colleagues. This was an important and interesting task, and a rare one in this case. Usually when you complete analyses on samples, just as in any field of science, you have a specific goal or idea in mind about what will be of importance in the sample. In the case of the WTC dust and smoke, I had no true hypothesis or "educated guess" about where to start the analyses, but we did know that these were massive buildings that had contained many business tools and possessions, they were constructed of a variety of building materials, and there was the intense fire. Furthermore, I wanted to make sure that we could eventually link the WTC dust measurements with any observed acute health effects, such as respiratory illnesses and abnormal child development, which required us to identify components of the WTC dust that could potentially cause long-term health effects. Therefore, even though we could not actually measure actual exposure because of the lack of personal monitoring equipment and a general plan, we might be able to provide information for reconstructing inhalation or nondietary ingestion exposures caused by at least the particles in the WTC dust.

The main materials of concern were combustion products, fibers from the many materials used indoors and the ionic and elemental materials in cement.

HISTORY LESSONS FOR THE WTC

We must remember that the collapse of the Twin Towers was not the first set of buildings to collapse because of an explosion. During World War II, buildings collapsed due to the bombing and firebombing of cities and industry with conventional weapons. Bombing was widespread over Europe and parts of Japan during the years of conflict. London suffered from continuous bombings during the war years. A quote from a book by my departed friend Dr. David Bates, who was an internationally recognized pulmonary physician and leader in the field of environmental health sciences, sheds some light on the times. His chapter on the Battle of Britain during 1940 was poignant. David was a corporal in the British Home Guard, and he kept a diary of the times. On September 7, 1940, he wrote:

> The sirens sounded in the late afternoon today, and some minutes after we saw a flight of 60 enemy Bombers with escort flying toward London. A few minutes later another flight of about 25 planes arrived. From then on

there was a continual procession of machines passing over going toward London. There were over 120 planes in one formation alone. Altogether we estimated there were 400 planes all going towards London. There was one continuous roar for over three quarters of an hour. Several large fires could be seen, casting a bright glow in the sky after dark. I was on patrol that night, and was walking along Alma Road at about 1 am, when I heard a plane overhead coming in my direction. I took very little notice since planes had been flying around the whole evening. A few seconds later I heard a whistle, and dived for cover. A bomb exploded quite near. The exposure was immediately followed by a second whistle, this time much nearer. The bomb exploded about fifty yards away. Three more fell soon afterwards. No-one was injured. I examined the crater when it was light, and would judge that the bomb near me was either a 150 or 250 pounder. The crater was about 20 feet across. (Bates 1997)

I saw David for the last time about two months before he died, just before I gave a talk in Vancouver, BC, on the aftermath of the attack on the WTC and the WTC dust. Although very ill his mind was still vibrant, and we talked about many things including my experiences in the WTC aftermath. The courage he displayed during his life was still there, and reinforced my need to move on, continue to contribute, realizing that men and women like him were members of the world's version of the "greatest generation."

With the dropping and detonation of the atom bombs and the total destruction of Hiroshima and Nagasaki, Japan, the world was presented an entirely different set of exposure conditions and long- and short-term health effects of radiation exposure. Many reports have been written about those attacks and their aftermath. A good early example of the health effects is "The Effects of Exposure to Atomic Bombs on Pregnancy Termination in Hiroshima and Nagasaki," completed by the National Academy of Sciences in 1956 (#461, Washington, DC). There have been many more.

Many survived all of these diverse attacks, and people did move on with their lives, as difficult as it may have been at the time. Anthologies present pictures of the widespread destruction and suffering that resulted during the defeat of the Nazis and Japan during World War II. These should be part of the background given for any lesson about the attacks on the WTC, the Pentagon, and Flight 93.

4

THE PATH TO RESEARCH AND SAMPLE ANALYSIS

Why did we bother to analyze the WTC dust samples? My answer is related to the nature of building materials used in the post–World War II period, and what I had learned over the years about the inhalation of aerosols in the aftermath of a major catastrophe. To me, the situation at Ground Zero seemed capable of causing, at a minimum, respiratory (lung) effects because the buildings totally disintegrated into dust. This is partially because synthetic materials were used during construction and were also used in the interior office designs and other spaces since the early 1970s. The burning of these materials would make the materials in the dust and smoke a complex mixture. As a result, the nature of the crushed building materials and "nonbuilding materials" would have been different, and the gaseous combustion products released by burning synthetic materials during the first few days of exposure during the aftermath would have been much different than found in the debris and fires during, for example, World War II. I must note, however, the nature of the gases released from the destruction of military targets could also have been very toxic.

To put the WTC dust into a broader perspective of harmful "dusty" situations, the soil distributed in the Midwest during the 1930s, which has come to be called "Black Blizzards," is a good example. These were thousand-foot-high dust storms that occurred throughout the Midwest, yielding the Dust Bowl. These dust storms were caused by the poor farming practices at that time. The material resuspended into the air was dirt, not man-made fibers or other synthetic materials. However, there was lots of material, and the dust persisted for years. On many occasions, inches of dust were found indoors, and people had to go on with their lives trying to see through the high levels of dust present indoors and outdoors. I wonder if anyone has ever looked back in time with fresh eyes to estimate the amount of lung damage caused by breathing the resuspended dust that was present in that area, both indoors and outdoors, for many years. Perhaps the historical records will provide some glimpse of the Dust Bowl respiratory issues and benchmarks for any long-term health consequences of breathing WTC dust. However, we must remember that the WTC dust was inhaled over a much shorter period of time. Respiratory effects of the Dust Bowl era may be a good research project; however, the health effects data available from that period of time would be sketchy at best.

There were practical reasons why the WTC dust needed to be analyzed in detail. During the weeks and months following the collapse of the towers, I received many phone calls about what was in the dust. These calls led to discussions on the presence and levels of WTC dust in the home, what could or could not be cleaned, what to do with rare books that had WTC dust on the surface, and what health issues could be of concern. For specific health questions I referred the callers to physicians. One key question that I was asked repeatedly was, Can the residential areas be cleaned by personal cleanup strategies, or do they need to be cleaned by a professional? We will discuss this question in much greater detail later in the book.

I tried to help, as best I could, given the initial lack of information. As time went on, however, my messages did become clearer. I have to say that although these phone conversations could last up to two hours, I found each call to be worth the time. In addition, I received e-mails detailing situations surrounding the caller and family or friends during the collapse, including the condition of residences or businesses.

In consultation with the other members of our sample analysis team, Cliff and I prepared to send out samples. We made the decision to have analyses completed on the WTC dust that would provide information on quite a few organic (carbon-based chemicals) and inorganic (elements from the periodic table and ionic species) materials. In fact, we decided that the analysis would be comprehensive. We were aware that, although different organizations would probably take dust samples and submit them for analysis (again, at the time we did not know of anyone else having samples other than the samples collected by NYU), each of these other organizations would focus on their individual analytical strengths or preconceived ideas about what was important. Therefore, by analyzing for as many components as possible, it was reasonable to assume that we would find the presence of anything of importance for testing exposure- or health-based hypotheses. This approach is useful when dealing with an unknown, but it was not truly a "shotgun" procedure, since, as mentioned before, there were basic chemical properties and specific classes of target compounds that would be associated with construction materials and fires. We also believed that a broad spectrum of results would be helpful in validating others' work on detecting specific agents. This approach was the right thing to do given the complexities that resulted in attempts for restoration during the aftermath. In the end, there were also some components of the initial WTC dust and smoke aerosol that we would never have any information about and that could never be reconstructed for use in exposure analyses.

Our final list of chemical components to be measured by the analyses included the physical and chemical nature of the WTC dust, the size of the dust particles, metals and other elements, anionic and cationic species, radioactivity, a general scan of organic compounds, semivolatile organic compounds, polycyclic aromatic hydrocarbons, dioxins and furans, polybrominated biphenol ethers, the pH, and corrosion properties of the total dust. There were lots of materials with various characteristics.

Each component of the WTC dust was selected for analysis for a specific reason. Based on past experiences with unknown dusts, we felt that morphology gave us our best chance to quickly define the overall characteristics of the particles. Information on the nature and origin of particles can be obtained from morphological analyses that tell us about their size, shape, and/or surface. We accomplished this by examining the particles using microscopy and analyzing pictures, while simultaneously collecting X-ray spectra of individual particles. The analyses identified the physical particle size range, and the particle shape and form. An experienced microscopist would provide clues as to the identity of the types of particles present, including minor and major contributors. The results helped support our initial decisions on what type of compounds or elements should be measured using our sophisticated chemical analytical tools. This process can be described as a forensic scan of the WTC dust to establish the direction of further analyses and the interpretation of the results in terms of exposure.

As I began contacting colleagues about the possibility of working with us on the analysis of the WTC dust, I had nothing to offer them except WTC dust and smoke samples. Each person mentioned offered to help as a volunteer. No one asked for or expected any compensation; each just had the desire to help.

A minor point, however, needs to be made here. Over the years accusations have been made about the analyses completed on WTC dust and ambient air samples. In terms of the work done by my colleagues within the NIEHS center's response to characterize exposures, they all did superb work. We could not do every analysis, since the material and the timing of our entry into the situation would not allow for better planning for the analyses to be done. It has been stated by others many times before "no good deed goes unpunished," and that statement was alive and well during the time I was involved with the 9/11 aftermath. The criticisms leveled at investigators and their analyses were unwarranted. All analyses on these three outdoor samples were done well and done as quickly as possible for a volunteer effort. No one directed or asked our team to complete or not complete any particular measurement.

Because of all those who graciously volunteered, it was easy to set up a WTC dust analysis program and get the three samples analyzed as quickly as possible. We had three outdoor samples of considerable weight, over a pound each. Therefore, enough material was available to be sent to everyone. Eventually, we were able to defray some of the analytical costs (e.g., supplies) for each investigator because of funding by the NIEHS. However, the time and effort spent on reviewing and interpreting the results obtained on the three WTC dust samples during this time period were still voluntary.

The techniques we used to analyze the samples included an array of state-of-the-art techniques available to investigators in the environmental health sciences and exposure science in 2001, and, for that matter, many other areas of science (e.g., analytical chemistry). Today, with the advances that have been made in analytical technology since that time we probably could and would take advantage of more tools for new samples.

The collaborators who agreed to analyze samples included Brian Buckley of EOHSI, who analyzed samples for metals and other elements using an Inductively coupled plasma mass spectrometer. Among the metals measured were the neurotoxicant lead and the cancer-causing compounds arsenic and cadmium.

The volatile organic compounds attached to the surface of WTC dust were measured by Cliff Weisel, including materials that would have been part of any unburned jet fuel. Polycyclic aromatic hydrocarbons were measured by Steven Eisenreich (Department of Environmental Science, Rutgers University). These classes of compounds are commonly emitted during combustion. Semivolatile compounds were measured by Brian Buckley. These compounds would be found in a wide range of potential source materials that included jet fuel, fuel oils stored at the WTC site, cleaning agents, pesticides, and combustion products.

Polychlorinated dioxins and furans, which are products of incomplete combustion (poorly controlled burning of materials), were measured by the U.S. EPA, Region 7 laboratory through the efforts of Dan Vallero, EPA. Brominated diphenol ethers, which are fire retardants, were measured by Dr. Robert Hale, College of William and Mary, Virginia. Each of the organic materials or classes mentioned here were all measured using a well-defined extraction procedure to clean up the samples and isolate the materials for analysis, and then by using a gas chromatograph and mass spectrometer (GC/MS) or liquid chromatography and mass spectrometer (LC/MS) instrument package.

Mass spectrometry is an analytical technique used for identification of chemical structures, determination of mixtures, and quantitative elemental analysis based on application of the instrument called a mass spectrometer (MS). It can measure the masses and relative concentrations of atoms, isotopes, and molecules by using basic magnetic forces on a moving charged particle to separate the charge particles by mass. It provides a kind of a fingerprint of a material by the mass of its parts. The coupling of the MS to a gas chromatograph (GC) or a liquid chromatograph (LC) allows for the separation of materials prior to entry into the MS, which allows for better quantification of the materials present in the sample. The combination of a mass spectrometer and a form of chromatography makes a powerful tool for the detection of trace quantities of environmental chemicals.

The morphological analyses of the particles were completed by Jim Millette, MVA. He used a combination of polarized light microscopy, scanning electron microscopy with energy dispersive spectrometry, and scanning electron transmission microscopy with energy dispersive spectrometry.

Aerodynamic particle size separation was completed by Lung Chi Chen (NYU). This process classifies the particles according to size accounting for physical shape and density, and also by the way the particles would behave when suspended in the air. He completed the ion chromatography analyses for acids and bases in the WTC dust and determined the pH of each sample. Later, he determined the pH of the different particles' size ranges.

Finally, we measured functional groups of individual compounds in the samples using a Fourier transform infrared spectrometry (Barbara Turpin, Rutgers University) and radioactivity. Gamma rays and elements from various radioactive series found in the periodic table were measured in the samples.

For those who are interested in analysis of environmental exposures, analytical chemistry, or forensics, details of each device and the quality control procedures used for each of the preceding analyses were documented in our 2002 article in *Environmental Health Perspectives* (see the bibliography). More details are found in the primary references provided for each technique. You can visualize a large number of people in laboratories across the eastern United States simultaneously processing samples, assembling data, and examining the results. We began the methodical process by weighing samples, recording a chain of custody, and packaging samples for shipment to each laboratory by the last week of September. Think of it as multiple "CSI" laboratories operating at once.

With our approach to analysis of the samples started, the next order of business was to get ready to obtain "indoor" WTC dust samples. There were many locations and many buildings that would take months to evaluate and then clean up, so it would seem that collection of indoor WTC dust would be an easy task to accomplish. Again, I was wrong—

this would prove to be a real challenge and one that almost became an obsession for Lung Chi Chen of NYU.

One piece of good news at this time was that things were moving quickly on the NIEHS front. On September 19, the director, Kenneth Olden, through Allen Dearry, had reached out to a number of NIEHS-sponsored centers to convene a meeting at Johns Hopkins University in Baltimore. The goal was to discuss planning and organizing an NIEHS response to the WTC tragedy. The topics on the agenda would be wide ranging and include exposure assessment, a longitudinal health study of workers, commuters and residents, and outreach efforts.

On September 20, *Business Week* published one of the first stories about what was to become at least two to three years' worth of articles published by the media titled, "What's Lurking in That Smoke?" The week before the *Los Angeles Times* and other news outlets had articles on the concerns about the reassurances from the federal government on the smoke, but a few of us were more cautious about the WTC dust. The articles had comments by Phil Landrigan of MSSM and me that related to the dust, and Phil made a strong point about his concerns for workers at Ground Zero.

Eventually, I was interviewed many times by the media, as well as others from the NIEHS Centers. One of the best interviewers was Anthony DePalma, then of the *New York Times*. As time went on, it became clear that science was being put on the back burner in favor of speculations and accusations. A bit more discovery and measured responses to questions and disclosure by all could have prevented some of the confusion and distrust that tainted rescue and recovery, and even the reentry and restoration as the WTC aftermath went on and on.

A MEETING IN BALTIMORE

The NIEHS meeting on September 21 had a large contingent of attendees from various NIEHS centers. Clearly, like many other WTC issues, time was of the essence if our group was going to have an opportunity to make a significant impact. We all had to make our way to Johns Hopkins in Baltimore by car or train, no planes were flying on that day. During the meeting it was clear that NIEHS wanted to put together a coherent unified response. I felt we were all a bit frustrated with the local officials by that time. However, the local problem would persist and get worse. The city of New York was in control, but it actually should have taken the offers of help from EPA much sooner.

The NIEHS centers included those from New York University Medical Center, EOHSI/UMDNJ/Rutgers University, Johns Hopkins University, and Children's Health Centers at both MSSM and Columbia University. NIEHS staff did begin to develop a plan of action. The three areas of most concern during the meeting were health effects, exposure assessment, and community outreach. Phil Landrigan was selected to coordinate the development of health studies, I was asked to coordinate the exposure assessment studies, and NYU and Columbia, which had already started community forums, led the outreach.

As a part of exposure science, the term *exposure assessment* can be simply defined as an approach to estimate the duration, frequency, intensity, and route of contact with hazardous materials. Exposure assessment is applied directly to estimate the risk of contracting a disease. The NIEHS exposure assessment group was given three main activities:

- to provide a systematic assessment and further research and analysis of the local population and worker contact as a result of the release of the dust and smoke from the attack on the WTC,
- to use all data collected by agencies and organizations to assess exposure for acute and long-term health effects, and
- to achieve collaboration and integration among the four centers involved and the NIEHS.

The NIEHS promised some funding to each center to complete each of the tasks defined at this meeting. As stated previously, some of these resources eventually helped defray some of the costs for the analyses of the original three WTC dust samples. However, it still did not nearly cover the expenses incurred by all for the outdoor dust analyses.

The exposure group activities were completed using the expertise at NYU Institute of Environmental Medicine for ambient air sampling for long-term outdoor exposure estimation; Columbia School of Public Health for data acquisition from many organizations and Geographic Information System (GIS) analyses; and Johns Hopkins School of Public Health for occupational exposure monitoring and characterization. Finally, there was our team at EOHSI who were completing WTC dust analyses and would eventually complete detailed exposure modeling of the plume caused by the fires. Many of these activities were coupled with health investigations among the workers and community. The key outcome was that the centers became a team under the umbrella of the NIEHS. This increased our visibility for becoming a point of reference for "hands-on" activities among ourselves and others.

As time went on, the NIEHS continued to supplement all of the centers financially, since the science and health problems just did not seem to go away. Actual research projects were eventually developed, and some were based on the initial voluntary efforts with the WTC dust samples. In the end EOHSI received between $750,000 and $1 million from NIEHS over the course of about three years to complete a wide variety of exposure and outreach tasks, and health-related studies. The other centers also received funding from NIEHS to complete a number of projects. Based upon a request from EPA that will be described later, Panos Georgopoulos and I received another $150,000 from EPA for another part of the exposure science issue, the reconstruction of fire and smoke plume, which we will discuss later. To this day, a number of the medical schools led by Mount Sinai School of Medicine, including Robert Wood Johnson Medical School, which sponsors EOHSI, receive funding (not from NIEHS) to care for the continuing needs of some of the workers and volunteers located in the New York and New Jersey area. In response to a question from me on the NIEHS efforts, Phil Landrigan said, "I thought NIEHS did a very good job considering that they had never done anything like this before. They moved quickly. They used clever mechanisms to get dollars out to the universities. They harnessed the best and the brightest. I thought they did much better after WTC than after Katrina" (Phil Landrigan, pers. comm.).

DUST AND OTHER ACTIVITIES DURING RECOVERY

During the remainder of 2001, I learned of other exposure science–related work being completed in Lower Manhattan. For example, measurements of particles being emitted from the remaining fires were being taken at Houston Street (about a mile north of Ground Zero) in October by the Department of Energy. The analyses were completed by Tom Cahill, University of California, at Davis. His work began primarily during the third week postattack, at which time measurements were being completed by EPA and the NIEHS Exposure Assessment team.

In mid-October, I was informed about an important exposure–health response study being started on firefighters by the FDNY through the efforts of David Prezant and his staff within the department headed by Kerry Kelly, Chief Medical Officer for FDNY. The study design intrigued me because the work appeared to be focusing on an important issue: the timing of the arrival of firefighters at Ground Zero. On October 16, 2001, Dr. Kelly was quoted in the *New York Times* article by Revkin as follows: "400 firefighters in the initial project are broken into four groups: those exposed to large amounts of smoke in the moments before the Towers collapsed, those who arrived after the collapse, and some who were on the scene in the first week, and a group who for various reasons had minimal exposure" (Revkin 2001). This proved to be a very wise decision. Their strategy eventually helped me to justify some of the theories we would be developing that related to the intensities and durations of contact with significant components and levels of the WTC dust to occupational exposures during the WTC aftermath. It has led to a number of productive years of collaborative research and friendship with David Prezant.

5

TIMING IS EVERYTHING
FOR EXPOSURE

As time passed, I spent countless hours trying to explain to the press and the public about the true nature of the exposure issues post-9/11. Part of the reason was the complexity of the issues for both the locations of concern and the time when the most important exposures actually occurred. This was due in part to the fact that we live in a "sound bite" world, and few people seem to take the time to listen to obtain a thorough explanation of a situation, and it seems to be a very real problem in the twenty-first century. I hope that the content of this book can stimulate you to learn more about science and math, and how the principles are used in understanding environmental events as well as to develop new products and tools for society.

Even today, the general public still does not know that the complex dust and smoke aerosol mixture initially suspended in the air during the hours after the collapse of the towers remains something of a scientific mystery. This mystery comes from the lack of data on the gases released during and just after the collapse of the towers, and exposures caused by the gases released from the fires. *The combustion and evaporated gases were never measured, and the names of the actual chemicals and their concentrations will remain unknown.* The gases emitted by the intense fires were not measured in Manhattan during the initial hours postattack. To be blunt, the instruments to measure the gases were just not present (a reasonable situation because of our pre–September 11 mind-set), and few, if any, would have been able to operate in that very thick dust-laden air within NYC. The reason for the latter was a lack of samplers designed to operate in high dust-laden environments. Therefore, we do not know what gases mixed with the dust and other particle-bound combustion products during the collapse and the period of time immediately after the collapse. As a result, we will never know how this complex mixture of gases and particles affected the exposures and specifically the lungs of rescue workers, fire-fighters, and others within the area around Ground Zero.

We do, however, understand part of the exposure problem based on our analysis of the WTC dust. This included the first twenty-four hours postcollapse, which eventually proved to be critical for thousands of volunteer workers. From 11 AM on 9/11 through at least noon the next day is what I now consider exposure period 1. (Remember this time as it comes back frequently in the story.) The major sampling data gaps occurred immediately postcollapse because of the lack of information about the levels of gases and the composition of the initial WTC dust released into the air. These gaps could not be filled by the limited measurements of asbestos fibers made during the first forty-eight hours postcollapse. Asbestos was one specific toxicant that could exhibit toxic effects. However, it usually takes long-duration asbestos exposures (i.e., ten to forty years) to yield health effects. In contrast, for gases emitted during a fire, the first effects could be observed only minutes to hours after exposure depending on the types of gases and the levels in the air.

Because of the initial focus on asbestos, here is some background on asbestos. It is a very toxic agent, which causes severe respiratory illnesses. One such illness is clinically defined as asbestosis. In some cases asbestosis leads to a cancer called mesothelioma. During the early 1970s, it was a major problem in New Jersey, because of the Johns-Manville Asbestos production plant in Manville. The long and very thin fibers were affecting both workers and their families. Asbestos was used for many years as a fire retardant and was heavily used as insulation material during World War II in the ship building industries. During the construction of the Twin Towers, it was applied to partially insulate one tower. This was the reason used for asbestos concern after the collapse.

However, the asbestos argument was not laid out properly at the time of the collapse. What most people did not know is that it typically takes years and long durations of contact with the material to see the expression of asbestos-related disease. So, asbestos was the wrong indicator for metric of exposure for WTC dust and its impact *on potential for acute (short-term) health outcomes.* This became a serious source of confusion since agencies and the media were making statements to the effect that the air was safe. How safe was it? In terms of ambient air levels of asbestos and personal exposure, they were correct—a point clearly made by EPA administrator Whitman on September 18. But what about anything else? This was an unknown, but at least the exposures diminished quickly in the ambient air. However, a full understanding of the complexity of the indoor problems that would face the residents, and which agencies would deal with as time went on, was not available.

THE TIME COURSE OF CONTACT AND EXPOSURE TO DUST AND OTHER MATERIALS

The issue of exposures that would have occurred over time after the collapse of the towers is important. Therefore, the postcollapse WTC events must be placed into a scientific and human time line. The time line will help define the contacts and exposures to contaminants in the dust that occurred postcollapse. While the situation at Ground Zero and in the surrounding area was evolving, the time line of exposure postcollapse of the World Trade Center Towers confused me for a while. So, I have placed a time line in the following pages hoping to use it to more accurately explain the WTC dust and suggest what it meant for cleanup and human health.

In contrast to many environmental events, the types of exposure that occurred over time during the aftermath were not all the same. Panos Georgopoulos and I described the types of individual population exposures across distinct time periods in a paper entitled "Anatomy of the Exposures That Occurred around the World Trade Center Site: 9-11 and Beyond" (see the bibliography). Most environmental releases have similar characteristics throughout the event, with the main change being the decrease in emissions over time. The WTC emissions did not behave that way. There were different scales of space and time, unknown and known gases and particles emitted from Ground Zero, changing emissions and sources, and a number of different exposure groups and activities. We determined five exposure periods, modified for presentation in this book (see table 5.1).

Table 5.1. Periods of the Time Course of Environmental and Occupational Exposures after the Attack on the WTC

Periods	Descriptors	**Pollutants**	Duration	Material
1	Generation of dust/smoke caused by fires and collapse of the Twin Towers.	**1. Combustion products gas/ particles.** **2. Coarse particles and evaporating gases from the collapse of towers.**	First few hours post collapse	1. Dust pulverization of building. 2. Jet fuel fires.
2	Continuation of initial jet fuel fires. Resuspension of settled dust/ smoke.	**1. Combustion products gas/ particles.** **2. Coarse particle resuspension.** **3. Evaporation of gases from piles.**	9/11 through 9/13	1. Settled dust. 2. Jet fuel and other burning materials.
3	Smoldering building fires. Resuspension of settled dust/ smoke.	**1. Combustion products gas and particles.** **2. Coarse particle resuspension.**	9/14 through 9/25	1. Dust. 2. Burning debris and fuels. 3. Diesel emissions.
4	Smoldering and spikes of pile fires. Removal of debris by trucks and construction equipment.	**1. Combustion products—gases and particles.**	9/25 through December 20	1. Burning debris and fuels. 2. Diesel emissions.
5	Indoor Environment	**1. Initial dust and smoke** **2. Smoldering fire plume entry into building** **3. Resuspension of dust and smoke (Building+ Ventilation)** **4. Diesel infiltration**	9/11 through 2003 (After EPA program ends)	1. Settled dust/ smoke. 2. Re-suspended dust. 3. Cleaning activities. 4. Fumes from diesel emissions.

Source: Adapted from Lioy presentations

Exposure Periods Post Collapse of the North and South Towers

Exposure Period 1

Exposure period 1 was the most intense time of exposure for any survivor in Lower Manhattan. It included the collapse of the towers and the next one to twelve hours. During the time just after the collapse, southern Manhattan was enveloped by an unprecedented high concentration of dust, smoke, and gases. (The graphic images of people covered with dust and debris have been shown on the covers of weekly news magazines, and most people who watched the commercial and cable news networks saw these images again and again. An example of people in downtown Manhattan covered with WTC dust is shown in figure 5.1.) Thus, period 1 resulted in a very intense period of uncharacterized particles and gases affecting residents, commuters, and rescue personnel in southern Manhattan, or to put it simply; all who were "in harm's way." As I stated earlier, Matt Lorber estimated the levels of particles during this period of time to be well above 5,000 µg/m³ of air. On a

Figure 5.1. Survivors of the collapse covered with WTC dust in the streets of southern Manhattan immediately after its release from the collapse of a tower. Associated Press Images/ © Gulnara Samoilova. Used with permission.

typical high-pollution day in New York City, particle levels may be *only* 40 to 50 µg/m³ of total mass.

As intense fires burned postcollapse, the disintegration of the towers left Lower Manhattan with an unprecedented situation. On the surface it would appear to be one usually not found outside very arid cities, but WTC dust was definitely not as simple a mixture as desert sand. We had dust/smoke from the Twin Towers depositing on the ground and surfaces both outside and inside offices, business, and residential buildings throughout southern Manhattan. Satellite images taken of southern Manhattan showed a significant whitish coating, like an early winter snowfall (see figure 2.1). Furthermore, no one knew what the deposited material was made up of at that time; therefore, the dust should have been approached with caution in areas with significant loadings on the ground or surfaces in buildings. Remember the first material sampled was only asbestos, and the levels of gases released by the fires would remain unknown.

The intense levels of emitted smoke and resuspended dust material continued for ten to twelve hours postcollapse. To further complicate the situation, WTC building #7 collapsed later in the afternoon on the 11th.

Images and news footage showed the dust being resuspended every time an emergency vehicle went through an area with WTC dust on the ground. People running from Ground Zero covered their face the best they could, but that was a difficult task, as seen in figure 5.1. The routes of exposure would include direct inhalation of gases and particles. Inhalation would occur because of dust resuspended from clothing and surfaces. Ingestion would occur because dust attached to the hands and face would be swallowed, or coughed up from the lungs and swallowed. Images of the dust were seen all over Manhattan. For instance, a friend reported that she was walking near Central Park where she remembers hearing the sirens from unmarked police vehicles as they were racing through the streets coming uptown, and they had WTC dust blowing off the vehicle surfaces.

At the same time, the smoke associated with the burning fires that intensified after the collapse of the WTC moved over most of Lower Manhattan and sections of Brooklyn. For most people, the plume was just a visual reminder of the tragedy, since the weather, as mentioned earlier, was such that the smoke from that very hot fire drifted high up over Lower Manhattan and Brooklyn. Therefore, the most intense levels of particle and gases from the burned debris went over the heads of the general public in Brooklyn and out to sea. There were places in Brooklyn that were impacted by the plume but had levels of particles and gases much lower than what would have been the case if the weather had produced stagnation in the city (no air movement) rather than a steady wind directed from northwest to southeast.

Exposure Period 2

During September 12, we defined the local situation in the New York metropolitan area as exposure period 2. The large and small dust particles deposited after the collapse were resuspended around southern Manhattan, and gases/fine particles were emitted from intense fires that continued to burn at Ground Zero. At this point, however, the people who were most directly affected had decreased in number. This group included the volunteer and uniformed rescue personnel, true heroes, and a number became wounded warriors.

A number of investigators collected settled total dust samples, which provided information on particle size and composition, but again not the gases contributing to human exposure during periods 1 and 2. The levels of WTC dust and fire smoke inhaled were still high but not as high as during exposure period 1. Exposures would have been primarily ingestion and inhalation of resuspended dust, and direct inhalation of combustion particles and gases from the fires.

Exposure period 2 lasted until Friday the 14th, and interpretation of the situation became complicated because of brief, but intense, plume impacts at various locations within and outside southern Manhattan during the afternoon and evening of the 12th. From about midmorning on the 12th until very early on the morning of the 13th, the winds shifted completely around the compass, and just about everyone within forty miles had an opportunity to smell the WTC fire plume for a short time. The WTC plume became a new focal point for the media and the general public, and I could no longer get most of the media to focus on the dust versus the plume. This situation lasted until about mid-November. During that time the concept of two separate types of WTC exposure events, WTC dust versus WTC plume, was difficult to explain to anyone.

On September 14, it rained, washing much of the resuspendable material from outdoor surfaces in downtown Manhattan. That weekend we also saw the indoor and outdoor cleanup of Wall Street with the expectation of getting back to business on Monday. I agreed with the idea, as an American, believing that our symbol of strength needed to be displayed. However, there was a downside: the rest of Lower Manhattan had many locations indoors that were not clean, and many would not be cleaned for months. Ground Zero was still in the rescue phase of activities, and there was an intense need to reinforce the use of respiratory protection within Ground Zero. Again, without good communication, there was a series of mixed messages being distributed to the public.

Exposure Period 3

September 15–25 marks exposure period 3. By the 16th, the fires had weakened and were limited to specific locations across the site. There were sixteen separate fires on the 17th. During this exposure period, parts of the area still had occasional instances of resuspended dust and smoldering fires, but these would yield much less intense inhalation exposures. This included exposures to all workers and returning residents. However, during that time a new element was added to the exposure equation: odors that infiltrated indoors. The odors were from the fires that burned since the beginning of the aftermath. Now that some businesses and agencies began to reoccupy buildings, the odors were noticed by workers in the buildings, and I received a number of calls about this problem. Decisions to stay or leave were at the corporate level or based on individual reactions to the smell. Thus, the concern continued to shift to the plume even though the fires were continually getting weaker.

EPA staff was given access at Ground Zero on September 17 to start sampling, after having to obtain approvals. They began limited sampling on the 21st. Their full air quality monitoring system was up and running on the 26th of September, which was impressive considering that the period of recovery had just started and bodies were being found each day.

Exposure Period 4

On September 25, the last appreciable rain fell in NYC for the next six to eight weeks. That marked the beginning of exposure period 4. It ended when the fires burned out, on December 20. There were some days of noticeable smoke plumes that emitted fine particles and gases. There was also debris removal (resuspendable particles) and transport of the debris to the Staten Island Landfill.

Outside Ground Zero, as reviewed by Phil Landrigan and confirmed by others at NYU Institute of Environmental Medicine and the EPA, the ambient air was returning to typical New York City levels of common air pollutants. The EPA was able to monitor the outdoor air quality during the entire exposure period of the aftermath.

Exposure Period 5

Exposure period 5 entails longer-term indoor deposits of WTC dust, which had been lurking on the sidelines since the day of the attack. Actually, this dust had been there since the collapse of the first building. The source of the exposures would be the dust that had settled within offices, businesses, and residential buildings. As expected, indoor WTC dust became a source of considerable concern to the local community as they tried to assess the damage and the viability of their homes and businesses, and the level of cleanup of building interiors. The different types and degrees of indoor cleanups and indoor and outdoor demolitions would fuel concerns about persistent WTC indoor dust and whether or not it could cause acute or chronic health outcomes upon returning to buildings. In his book *Fallout*, Juan Gonzalez made a good point: there were no good interim metrics available for assessing the indoor cleanup. The 1 percent asbestos per unit mass of all material was just guidance for cleanup, not a health-based standard for dust deposited indoors. However, the development of surface dust standards or guidelines for contaminated surfaces remains an unresolved issue in 2009.

The fact that in many cases there were inches of dust clearly indicated the need for professional cleanup. I think this point was a constant problem in the messaging from and between various agencies and other organizations. Eventually, EPA did the right thing by quickly developing good cleanup metrics and protocols when it took over responsibility from the city in early 2002. In 2008, there were still a handful of buildings that remained closed, and one, the Deutsche Bank on Liberty Street, was still being deconstructed brick by brick; however, most buildings have been cleaned.

INDOOR SAMPLING

It was clear to me well before the meeting on the 21st in Baltimore that the indoor exposure and cleanup would be serious issues and extend exposure period 5. We began to make plans to get samples. Lung

Chi Chen of NYU and I started to map out a plan to get indoor dust samples. We both had retrieved outdoor samples and thought that getting indoor samples would be a simple task. What a naive thought that turned out to be! To this day, no reasons have been provided that explained the obstacles that we encountered during our effort to secure indoor dust samples.

Despite considerable debate and published advisories, there was no agreement on how to effectively determine the best way to eliminate the WTC dust that persisted indoors during the days and weeks after the collapse. Some guidance documents and web addresses tried to provide needed information, but they were incomplete and, at times, confusing. The New York City Department of Health appeared to have the authority, as well as FEMA. However, neither truly took charge, and I was amazed that the EPA took control and addressed the indoor issue, even though it did not have well-defined regulatory jurisdiction for indoors. In retrospect, this important scientific and engineering decision was the correct one for EPA, but in the end they were unjustly chastised by the inspector general, the public, and the media. Today, the EPA National Center for Homeland Security appears to have much more authority for remediation post terrorist events. However, through 2008, indoor surface cleanup values for dust were still hard to find.

The EPA, however, has made strides with the development of provisional advisory levels (PALs) for air and water over the past four years. PALs are threshold exposure limits for the general public, including susceptible and sensitive subpopulations. This tiered set of values is used in conducting threat scenario (hypothetical events) health risk assessments and for developing risk-based cleanup levels that assist with the return to normal operations. A good start, but they still would not be of any value in a settled-dust-type event. When will settled-dust PALs be completed, and for what substances? I have asked that question more than once of EPA scientists at their center (formed after 9/11), but I was never given a meaningful answer. The gap needs to be filled for PALs in units of mass per unit area of surface covered for toxic substances that settle indoor. Such standards or guidelines would provide a sound basis for starting reconstruction after the period of rescue and recovery, and will be illustrated for lead found in the WTC dust. However, it should be noted that we have had surface loading standards for radiation for years, and it should be a simple task to translate that approach to PALs for deposited chemical agents.

The indoor issue was a very tough problem to deal with, and there will be quite a bit more on indoor cleanup and exposure in the following chapters. Exposure period 5 lasted a very long time, but after the EPA management made its fateful decision to ignore the advice of a panel of experts in 2006, I finally turned the page on WTC, and focused more on the critical issues that still exist as unfinished business for homeland security and disaster response in general.

6

A SCIENTIFIC FRAMEWORK

Soon after our meeting on September 21, there was a flurry of activity among the NIEHS Centers to start implementing the three components of the NIEHS health response plan to the attack on the WTC. One of the major opportunities provided by the NIEHS was allowing us the ability to request data and information from various agencies, and to begin to get the approvals necessary to access particular groups of workers and the general public.

At the start of exposure period 4, on September 26, there was a distinct ramp-up in actions by agencies, such as the EPA, in the WTC area. Because most of the activities involved measurements in the air around Ground Zero, outdoor air quality became the major topic of attention. Again, however, this focus tended to confuse the issues since the concerns about the WTC dust were now mixed in with concerns about the general air quality of New York City. From September 26 through December 20, 2001 (period 4), a wealth of ambient gaseous chemical and fine-particle air quality data were collected by EPA, the NIEHS investigators, and others. Prior to this time, only a few volatile organic compound (VOC) measurements (e.g., benzene and toluene) had been made by USEPA Hazmat and OSHA between September 13 and September 25 (period 3).

Even today, the question still remains: how do we avoid having such significant and important gaps in data and agents of concern in the beginning of a sudden attack or other disaster? This point was discussed at length in an article I published with Pellizzari and Prezant in 2006 in the major scientific journal *Environmental Science and Technology* (see the bibliography). It was on the lessons that *still* need to be learned from 9/11, but no one commented on the topic until about six months later! Then it only received a better review after a strong plea by Dr. Charles Rodes of Research Triangle Institute in North Carolina for a special panel session at the 2007 International Society of Exposure Science Annual Meeting in Durham, North Carolina. It is a pity, but you could see that members of the scientific community were moving on or drifting backward toward more typical topics of research, such as children's exposure to pesticides. Each is important, but not as important as the ultimate safety of all children.

A 2008 report by the GAO (see the bibliography) pointed to similar concerns. In addition, even before such data acquisition and analysis is started, realistic rather than politically driven policies and procedures need to be in place to protect workers, volunteers, and residents—whether that be exposure avoidance or respiratory protection in the affected area(s). I think that members of the emergency response community understand, but I do not know why they continue to tolerate the old-style respirators. Someone needs to answer that question and change the way we operate in emergency response situations not requiring air packs.

The general public, however, is still basically clueless about what to do during and after a chemical, radiological, or biological event. I am sure that if something occurs they will run out in the street to see what happened, as is the natural tendency of people, hoping that they are just a viewer of the disaster, or can get a video to sell to news outlets. A clear mistake! When was the last national emergency response quiz taken by the general public? I truly do not remember. Have we ever been graded? Never! I will discuss this point in more detail toward the end of the book.

COLLECTING PARTICLES, BUT THE RIGHT ONES?

By the 26th, people were beginning to take thousands of outdoor data points that marked the beginning of exposure period 4, and the measurements continued for many months. At a later time, EPA would state that it took over ten thousand data points. But were these the correct compounds? For the period in question, probably most of the gases were correct. During exposure period 4, the EPA and others were primarily taking fine-particle samples. Under normal circumstances, the measurement of fine particles would not have been given a second thought. However, in this case the particles present in the air were from the smoldering fires, and some were those large particles found in the WTC dust. The fires at Ground Zero were uncontrolled, and larger particles probably were present, but the sampling methods available to measure fine particles were not designed to collect larger particles. They were no longer of concern to the environmental community. The samplers were called "Hi-Volume" samplers; they had a peaked rooftop and basically collected everything in the air. Even EPA could not find one, a point that has been mentioned more than once to me by Dan Vallero. Today, we still cannot effectively collect a large sample of particles in the range between fine particles (those between 2.5 micrometers in diameter) and particles 10 μm in diameter. However, these sizes are well below the range of particles eventually measured in the WTC dust. Hi-Volume samplers are still not readily available, and they still have no regulatory purpose because of the lack of a PAL! The result is a complete lack of data on particles of larger size, the supercoarse ones, that could be suspended in the air for periods of time in Lower Manhattan or another city if a similar event occurred today.

Just to give you the mind-set at the time, EPA did collect dust samples, but they only submitted those less than 44 μm in diameter for analysis (keep that term, 44 μm in diameter, handy for later comparisons). Unfortunately, this would end up representing only a small fraction of the actual mass.

As for our own samples, approximately ten laboratories were analyzing portions of the three samples. For the initial interpretation of the settled dust, the best we could do was put the concentrations into a form that would give the quantity of each material per gram of total collected dust.

Since we did not detect the material in the air with air samplers, the results could not be placed in the usual form of the number of grams or micrograms (1 millionth of a gram) within every cubic meter of air sampled by a monitor that pulled the air through a pump. The devices for sample collection just were not available. So even if we could get access to Ground Zero during our days of sampling, there were no air monitors that could be easily used for such sampling. None were being built for such purposes since the country had effectively controlled the emissions of most large particles from industrial sources in the 1970s. In addition, there was no electricity. Thus, all measurement devices for coarse particles had been retired, and EPA could not find any for use even during exposure period 4. The situation still remains poorly addressed today, more than six years since the recommendations to add such monitors by Dr. Gochfeld and me (see the bibliography).

The reason we had moved as a country to measuring only fine particles was that since the mid-1980s, the most toxic effects of particulate matter were determined to be associated with the fine particles (i.e., those less than 2.5 µm in diameter). Furthermore, because of the success of stiff EPA regulations in the early 1970s, we never expected to have to deal with very high concentrations of large, supercoarse particles in the air again. As a consequence, we shifted to measuring the very small particles since our air is generally cleared of large particles in most urban and suburban communities. We worry about the very small particles because of their effect on the lungs and now the heart.

Based on the three WTC dust samples we collected and the dusty situations shown on TV, it was clear that the depth of the settled WTC dust was quite variable outdoors. Therefore, we also could not easily calculate another typical measured quantity: the number of grams of a material per square foot of dust on a surface. Either the samples were not taken on flat surfaces, or the amount on a surface was just too deep to obtain the entire sample from the surface. Normally, to sample surfaces, one would need an exact template of the area sampled, and all the dust would collect within that area. However, one could use these results to project the amount per square foot when a person weighed the amount of WTC dust per square foot present on flat surfaces in a home or office. This was not easily accomplished due to the conditions that could be encountered. Our results, however, could be used to provide some level of information to begin to estimate the dirtiness or cleanliness of a building interior and how to approach cleanup and restoration and, in the end, rehabilitation.

I mentioned before that the indoor cleanup issue would eventually begin to take over exposure period 5, and that began to happen by the

end of September. At that time, the indoor problem began to be identified because of the need to have professional cleanup of businesses and
residences. Not only would the interiors of buildings have to be cleaned,
but the ventilation systems of all buildings would need to be checked
for WTC dust that entered the intake and exhaust systems as the dust
spread during the minutes after the collapse. Window air conditioners
also needed to be checked. As with the respirator issue, guidance was
being given by the city and others about cleaning up the interiors, but
again it seemed like people were going their own way through contractors hired by residences or their insurance companies, and personal
approaches to clearing the dust from their homes and businesses. Some
of the guidance was not adequate or contradictory. (A discussion of the
messaging on dust cleanup is presented in the 2003 GAO report.)

Unfortunately, the EPA became the whipping boy for the media and
the community. There were, however, plenty of other groups that did
not step up to the plate at all on the indoor issues. For example, FEMA
continued to get a pass through 2008. Once the EPA did take control of
indoor cleanup, lead in part by the efforts of the deputy administrator
of Region 2, Kathy Callahan, I felt that they did a very credible job in
attempting to come up with a plan for cleanup that was based on sound
engineering and science. However, by that time few were listening, and
most were critical of everything that was being considered or done. The
early delays caused by the lack of a coherent interagency plan were a
major part of the problem.

Another example of the dust confusion and the "who's in charge
problem" was how contaminated cars were handled after 9/11. It took
months to sort that issue out, but eventually EPA condemned all vehicles. The dust would be present in all plush surfaces including rugs
and seats, and even detailing of the vehicles would be unsuccessful. This
was a wise decision by EPA.

In the beginning of October, at the time the NIEHS researchers were
organizing their response, the Agency for Toxic Substances and Disease
Registry (ATSDR), in conjunction with the NYC Health Department,
designed an indoor dust study. Why weren't the NIEHS exposure group
investigators given any information on the design or on the questions
being addressed by the sampling study? I never did get a satisfactory
answer to that question. At the same time, Lung Chi was finding it

increasingly difficult to get access to downtown for indoor sampling
even though we were scheduled to sample one of NYU's buildings!
The indoor issue became another missed opportunity by the city and
ATSDR especially because we were beginning to learn a lot about the
composition of the dust. The development of a coordinated effort with
the NIEHS Exposure Assessment Group could have yielded a much
more comprehensive set of analyses on the ATSDR samples.

Private contractors were having better luck getting access, but their results would remain privileged information even though this was an act of
war on America. The fear of litigation became a large issue over time, and
unfortunately litigation would become another aspect of the aftermath.

People wanted answers, but no one could provide a satisfactory set
of answers. The issues were to become tougher to deal with as people
wanted to return downtown, but our knowledge was still very limited
about how to aid the efforts for reentry, restoration, and rehabitation.

One of the tasks for the NIEHS Centers Exposure group was to
examine occupational issues related to dust and smoke exposure while
working on the pile during the recovery effort. Since we were organized
after the rescue period, looking at prospective occupational health issues during recovery was the best that could be done. The research was
led by Dr. Alison Geyh of the Johns Hopkins School of Public Health.
As was the case for the rest of us, she did not have an easy task getting
to Ground Zero and taking measurements of personal exposures among
the truckers at locations within the Ground Zero pile. During a conversation, she indicated that it was very difficult to get access to the workers, and at times she needed support from the truckers to take *personal
monitoring* samples, a problem common to all. Her work started on
October 1 and continued for a month, and for her team it was a series
of twenty-hour days. Alison said to me, "There were respirators, but the
majority wore them around their neck, not on their face, and they didn't
have the necessary training." She returned in April 2002 for a follow-up,
and at that time the exposures were low, but the concerns about health
at that time were high. Her measurements among members of the International Brotherhood of Truckers did show higher exposures for WTC
dust, and it was mixed in with fine particles released by the continuing
fires at Ground Zero. Alison measured particles, as well as airborne asbestos and volatile organic compounds, directly on the disaster site. She

also measured particles in the different sizes that workers were exposed to and found exposures to the large particles, as well as small particles. Alison did say that since her work started at least a month out from the 11th (exposure period 4), the fires were driving the exposure on some days, and that was when the fine particles were high. On other days, the exposures to particles were driven by debris removal activities, which would represent particles similar in character to the original WTC dust. The highest levels that she detected reached above 1,000 $\mu g/m^3$ in the pile and were above 350 $\mu g/m^3$ at the perimeter. Since the truckers would move around the site, the personal exposures of the truckers who did not wear respirators were a bit lower but still reached almost 200 $\mu g/m^3$ of air. If this were just the typical occupational nuisance dust, all the levels would have been below the accepted standard. However, WTC dust was not typical nuisance dust. An interesting point was that the volatile organic compound levels were low, lower than I would have expected, indicating less combustible material (e.g., fuels) to burn.

During the time Alison was starting her study, we were still waiting for the analytical results from the WTC dust samples. Then on October 2, I began to have conversations with Dr. Gary Foley of the EPA. He was the director of the National Exposure Research Laboratory (NERL) at that time. We talked about the long-term exposures that might need to be estimated for the WTC attack. He wanted us to quantify exposures from 9/11 through at least the first thirty to forty days postattack. Over the years we had been supported by NERL though a University Partnership agreement to develop a modeling framework that was new within the environmental health scientific community. Our goal was to be able to estimate the contribution of a toxicant in a human from the source of the toxicant though the environment and contact with a person, or persons, to the doses presented to the body or one or more biological systems (e.g., circulatory and respiratory).

It was clear that Gary wanted us to complete the following task: Reconstruct the plume emitted from the burning fires from the WTC, and provide a template that could be useful in assessing human exposure for epidemiological studies and risk assessments. That request was to be handled by Panos Georgopoulos's laboratory, and Panos and I spent a great deal of time planning the approach to the problem. The good news was that at EOHSI we had been developing a source to dose modeling framework called the Modeling Environment for Total Risk (MENTOR), which had components that could be used to tackle the problem. In the early 1990s I observed that for every environmental problem we developed a "new" model to explain the situation or the impact. What a waste of time and resources. So I convinced Dr. Bernard Goldstein, the director of EOHSI, that we needed to develop a research tool to examine the relationship between the release of a toxic agent for any type of source and then build a modeling system to follow the agent from the source into the human body. Not just a model of individual components, but a system that can incorporate individual models. We recruited Panos Georgopoulos, who was at the California Institute of Technology, and he was given the above task. Obviously, he succeeded in developing a framework to accomplish the goal that is now called MENTOR. I was pleased that we could put MENTOR to effective use in the aftermath, and now we apply it to many homeland security problems.

POSTEXPOSURE PREGNANCY PROFILES

We got one request early on for the use of our estimates from epidemiologists. That was from Dr. Mary Wolff of Mount Sinai, who was at one of the collaborating NIEHS centers. She wanted us to develop exposure profiles for about 187 women who were pregnant at the time of the collapse and were in proximity to the collapse for some period of time through early October. The idea had merit, and we were able to complete the project. A diary had been kept by each woman in the study. We used these to characterize each woman's location in space and time with respect to the dust cloud and the resulting plume through the second week of October 2001. These data were linked to information on each woman's location with respect to the dust plume after the collapse, their entry and activities in uncleaned buildings, and the location and relative intensity of the plume during the weeks that followed the collapse of the towers. The plume intensity was developed by estimating the relative surface intensity of the plume at an individual location for each eight-hour period after the collapse. Figure 6.1 shows an example from 9/11 of the mapped plume and the relative intensity of the plume in each grid. We were able to use this to estimate a relative intensity of contact with the dust or the plume for each woman and her location on each day. We adjusted level of contact since the intensity of the plume decreased from 9/11 to October.

Based on the information in these women's questionnaires, we adjusted the intensity of contact with WTC dust for time spent in buildings with WTC dust. For perspective, the lighter shading in figure 6.2 gives a relative indication of the potential for higher values, but since actual measurements were not available during periods 1 and 2, we could not easily verify the actual toxic levels. In the end, estimation of exposure was based on the location day or hour after the event when the plume

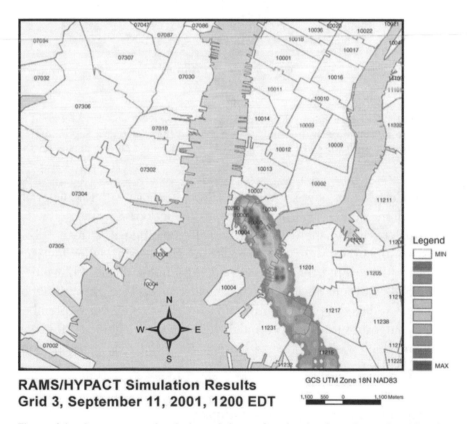

RAMS/HYPACT Simulation Results
Grid 3, September 11, 2001, 1200 EDT

Figure 6.1. A computer simulation of the surface levels of smoke emitted by the WTC plume during the afternoon of September 11, 2001. Courtesy of Dr. Panos Georgopoulos of the Computational Chemodynamics Laboratory of EOHSI Exposure Science Division and of Dr. Paul Lioy.

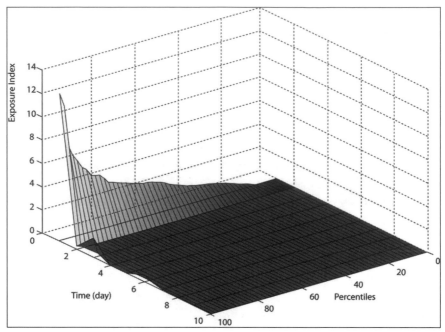

Figure 6.2. Exposure index calculated for pregnant women who were exposed to WTC dust and plume over time and space in the first thirty days past September 11. Derived from research completed by Paul Lioy and Panos Georgopoulos on the WTC aftermath.

was modeled. Thus, it was a representation of contact with the materials in the plume outdoors or materials deposited indoors.

Using the information for each day and coupling it with the diary of each woman, we were able to construct an exposure index for each woman that covered the entire period. The index for all 187 women examined by Mount Sinai (figure 6.2) clearly shows that the intensity of exposure reduced rapidly after the first few days (time 0 is September 11). Again, exposure periods 1 and 2 were the most significant for each woman, which was consistent with exposures for all emergency personnel and volunteers.

CREATION OF THE PLUME MODEL

My expectations were exceeded when Panos and his team were able to construct the previously shown three-dimensional time-varying simulation of the plume. However, all of these activities did take some time. The main purpose of the plume simulation was a mapping of the intensity (relative terms) of the plume as it passed over various locations in space and time, which was of value in describing the impact of the plume in many locations in Lower Manhattan. These were linked to available data from the ambient monitoring studies of EPA and the NIEHS centers, and the satellite images to check for prediction performance of the model on specific dates.

In addition to the plume model, Gary also wanted to use the wind tunnel associated with his laboratory to reconstruct the plume. This started with making a replica of southern Manhattan on a turntable in the wind tunnel, and then simulating the smoke releases from the sixteen-acre WTC site. They chose to simulate September 17. The work was completed in North Carolina, and I got a chance to review it during the development and the testing phase. It became a great source of data and photo-ops. However, the use of the data was limited. I do not know why, the collaboration just fizzled as people went back to their typical daily activities. This was a pity, since the information still has immense value for identifying the types of locations that could lead to toxic exposures in southern Manhattan for a variety of situations (e.g., accidents, terrorist chemical releases).

By October 4, we still did not have any actual results for the WTC dust, though my colleagues were hard at work finishing their analyses. It was a great effort that would come to fruition quickly. As you will see, the need for results suddenly became important.

During early October, physicians from Mount Sinai, Phil Landrigan and Steve Levin, and from the NYFD, David Prezant, began to report the incidence of a newly reported health outcome, which came to be called the "World Trade Center cough." Suddenly, a health issue put reality to the notion that the WTC dust may in fact lead to some immediate health consequences within members of the workforce and the general population, not just the long-term potential of the consequences of asbestos exposure. The acute respiratory and gastric health effects would evolve as a scientific and medical problem over time for certain groups of workers at Ground Zero, especially those who did not wear respirators, and the number of affected eventually rose to over six thousand. The analysis of the exposure effects issue took time, and the results from our dust analyses would become an important component of the discussion about the acute and persistent health effects.

STILL NO ACCESS TO INDOOR LOCATIONS

On October 2, a field team sent to downtown Manhattan by Dr. Chen to collect indoor samples was turned away by local security. The battle for indoor samples began in earnest. By October 4, we had sent e-mails to the EPA through the NIEHS management, but we still had no authorization to get back in the WTC area and take samples. On the 8th, Lung Chi contacted the EPA folks in the field at Ground Zero to see whether they could get us access. The stone wall was beginning to be built, but we were not sure by whom. Some resident associations were interested and contacted Lung Chi, but they could not grant us reentry to the WTC area.

ANOTHER TERRORIST EVENT

While we were waiting for data and trying to get to indoor locations, the country was rocked by growing reports of a second terrorist event, anthrax. So, not only were we dealing with potential acute health problems from the dust, but the country was beginning to see another acute exposure issue evolve, that of dermal and inhalation contact with a spore that truly can be called a weapon of mass destruction. The story began to slowly evolve in late September 2001 and ran in parallel with the WTC aftermath through December 7. Clearly, if not for the attack on the WTC, anthrax would have been the central theme on TV and radio concerning the exposures and health effects that could result from many types of terrorist attacks. The general public was experiencing overload, but we all still had to deal with the implications of these multiple anthrax attacks by a person who remained unknown until a potential suspect committed suicide in August 2008. Fortunately, the cost of human life was relatively low, five Americans, but a U.S. Senate office building was shut down, postal deliveries were disrupted, and the costs of cleanup of the Senate building and post offices in Washington, DC, and in Hamilton, New Jersey, amounted to approximately $100 million. We learned how envelopes of white dust could be used to effectively terrorize the general public. The fear illustrated another example of the true definition of a terrorist event. As a microbiologist, Jeanie eventually taught a number of superb courses directed at emergency workers and other professionals on the biological agents that could be used by terrorists, or generally as weapons of mass destruction (e.g., *pneumonic plague*). I wish there had been more opportunities for her to give a layperson's version of these lectures. They could help fill in gaps of knowledge that the general public still has about biological weapons and biological materials in general, and she could fill the room with listeners.

Eventually, I was asked to be part of a committee put together by the New Jersey Department of Health and Senior Services, led by then-deputy-commissioner Dr. Eddy Bresnitz. We reviewed the plans and made recommendations about the adequacy of the cleanup of the Hamilton post office before the postal workers could return to work. I remember standing with then-senator John Corzine of New Jersey at Anthrax Ground Zero on the day it was announced that the Hamilton Post Office would be reopening, two Ground Zeros for a person who never thought about terrorism before 2001.

The most upsetting part of the event for me was the day we received some mail at our home in a plastic bag that was labeled as having been irradiated for anthrax spores and, therefore, presumed safe. This envelope provided another graphic example of the signs of the times, and a visual representation of a terrorist act. However, times do change, and I wonder how many office workers or members of the general public would know what to do if they opened up an envelope today that contained white powder? Think about it, would you?

7

THE PRESENCE OF LEAD

On October 10, I received our first set of data from the WTC dust sample analyses, including metals and trace elements present in each of the three WTC dust samples. Our collaborators from NYU also measured metals levels in their dust samples taken near Ground Zero. This allowed us the opportunity to check the validity of one another's analyses and data sets, and characterize a larger picture. For the three dust samples that Cliff and I prepared, and subjected to multiple analyses, about twenty-six metals were measured, which will be discussed in the next chapter. However, one metal stood out at that time: the lead (Pb) levels in the dust, which was present in the paint used to coat metal surfaces in the towers. A micrograph and X-ray spectrum of a lead paint particle present in one of our WTC dust samples is shown as figures 7.1a and 7.1b, respectively.

To keep to the time line of the WTC dust events, I decided to discuss the lead now instead of waiting until the next chapter. The range in lead values found in all samples was not remarkable, but some values were over 400 µg/g of dust, the standard for the level of lead allowed in residential soil. However, what was more important was just the presence of the lead itself in the WTC dust. Since under normal conditions the concentrations of lead found in the dust were only slightly above soil residential standards, the New York City Health Department and others told

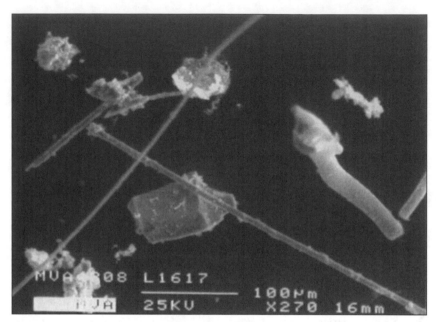

Figure 7.1a. Lead particle present in the WTC dust samples collected by Lioy and Weisel. Taken by MVA Project 4808, MVA Scientific Consultants. Used with permission.

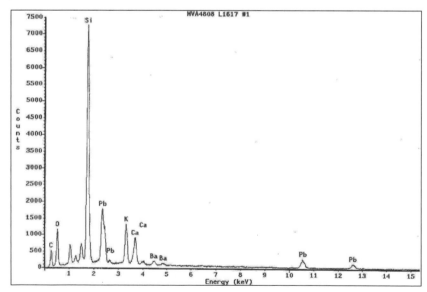

Figure 7.1b. X-ray spectrum identifying the lead particle and other material in the WTC dust. Taken by MVA Project 4808, MVA Scientific Consultants. Used with permission.

me that they did not consider lead to be a problem. I agreed that under "normal" or typical circumstances within the city, these concentrations would not be a problem. Before I show the fallacy of the argument, let me try to put lead in the context of environmental health problems.

Lead is a metal that is a well-known toxicant that affects children's neurodevelopment, including decreases in IQ, at relatively low levels in the blood. Prior to 1975, lead was commonly released as an air pollutant from the tailpipes of automobiles throughout the country. During that time, it was used as an additive to gasoline and was emitted intact from a car's tailpipe after the gasoline was burned in the engine. After 1975, lead started to be removed from gasoline and was banned in gasoline during the 1980s. Thus, lead was an air pollutant. However, lead also was deposited from the tailpipe on or near streets and can be removed from the air by rain. It also became a part of the storm water runoff from gasoline stations and distribution centers, and was released during refueling spills for gasoline pumps and cars. As a result, lead levels increased in urban soils of many cities, including New York City.

Lead was and still is an important public health issue for aggregate (multiple pathways for getting into the body) exposures from many sources over time. Of course we cannot forget the influence of lead-based paint on exposure among children, especially those who have significant mouthing behavior called pica. This behavior, involving eating unusual materials (e.g., paint, clay, etc.) has been noted in both primitive and modernized culture and can occur among all segments of the general population. In humans, we define pica by the material, in this case lead, that is ingested by an individual or individuals.

My concerns about the lead in the WTC dust were partly due to the fact that NYC officials were not looking at the full picture. Yes, there was lead in city park soils well above 400 ppm. So, they were correct in stating that the concentrations were not as high as in other locations, but the issue could not end there. As explained in conversations with the EPA (set up by Gary Foley) and NIEHS managers, my concerns were that the situation people would encounter when returning to their homes would not be just the concentration of lead present in the dust, which we all agreed was only slightly higher than the soil standard; it would be the amount of dust in the homes that contained these lead concentrations. This would result in dust and lead loadings on accessible surfaces that are much higher than what a person would normally find in the home, and higher than the standard of 40 µg of lead per square foot in residential settings.

Thus, to begin to deal with the problems of WTC dust, one could not just worry about the concentration of a toxicant in the dust, but the total amount that would be present on surfaces that are accessible to a child or adult (available for contact). We all need to remember that some residences and buildings had layers of WTC dust, not just a thin surface coating, which accounted for my specific concerns. Our discussion on the evidence for this exposure issue led to the following conclusion by NIEHS (excerpt of e-mail from Ken Olden, through Dr. Wheeler, to author on October 15, 2001):

> Lead Levels in Dust from the WTC
> Rutgers University's (and UMDNJ Robert Wood Johnson Medical School's) Environmental and Occupational Health Sciences Institute (EOHSI) has analyzed outdoor samples around the WTC and found concentrations of lead ranging from 103 to 635 micrograms of lead per gram of dust. New York University analyzed separate outdoor samples with lead concentrations ranging from 38 to 330 micrograms of lead per gram of dust. Because of the collapse of the WTC many of the homes in the area sampled already have significantly larger amounts of dust (perhaps several inches deep) as compared to very small amounts envisioned by the HUD standard, which assumes there is only one gram of dust per square foot. Therefore, if the dust in neighboring residences contains similar concentrations of lead to those samples collected by NYU and Rutgers, there may be significant amounts of lead in those residences that could pose a significant health threat.

This conclusion led to a public statement by Tommy Thompson, secretary of Health and Human Services at the time, on the lead issue. The WTC dust issue finally had a focus other than asbestos, and that was important in relaying accurate messages about the nature of the indoor issues in a scientifically sound manner. Lead is lead. Unlike asbestos, there were no concerns about fiber length or width and form. The duration of exposure to lead and the consequential effects were not twenty to thirty years in the future, which I thought was important for defining cleanup strategies and implementation plans. Lead could cause significant health effects in children (learning disabilities, IQ loss, etc.) and adults (hypertension) over a much shorter period of time. Lead also could be used as a way to separate the indoor dust from outdoor air quality issues, which found the outdoor levels of various pollutants, including lead, returning toward normal levels.

The lead issue does provide the window into the current need for PALs for settled dust indoors. For example, what if the compound with the highest loading in residences was a toxicant like cadmium, arsenic, or any number of organic compounds? Could reasonable and sound advice be given to a community or a municipal government quickly? In cases where sediment is deposited in an area after a flood when that area is near a known hazardous waste site of a large industrial facility, we cannot easily provide information or guidance on the ability to go in and start cleanup and restoration. So, settled dust or dried sediment or mud needs PALs in short order. From a 2008 GAO report, it appears that the EPA is at least considering a PAL for asbestos in settled dust, but that is insufficient. We need PALs for dust and the toxicants that can be present in dust.

PUBLIC DISCOURSE ON LEAD AND RESEARCH: A MEETING ON OCTOBER 18

On October 16, the *New York Times* published an article called "After Attacks, Studies of Dust and Its Effects" by Andrew Revkin and was reported on page 1 in the Science Times. Although it did not contain a lot of information, at least the article started to set the stage for discussion of the complex problems that lay ahead, including dust versus air quality, indoor versus outdoor cleanup, asbestos versus acute respiratory effects, community versus workers, and confusion versus logic. One of my comments reported in the article was "If I were a member of the general public, I would probably be scratching my head by now." Clearly, there were many mixed messages and a lack of information, even five weeks post-9/11. The asbestos issue still dominated the discussion at the expense of the workers' immediate health issues caused by the lack of respirator use, and the need for decisions on how to approach indoor cleanup. In the same issue of the Science Times, October 16, there was a page 1 article on *anthrax*.

The NIEHS centers were planning a meeting with the general public on October 18 in downtown NYC. We were just beginning to receive analytical data, and I barely had time to look at the metals results, let alone begin to understand this small window into what was in the dust. I found myself being uneasy. The residents wanted answers, but did we have them? No! The meeting was put together by George Thurston of NYU and his staff and Columbia University for the NIEHS, and I commend their efforts in putting this meeting together during such a difficult time. It was developed with great expectations from the local residents, and the room was filled with about four hundred people. The presenters were from multiple organizations, mostly from the NIEHS Centers in the New York metropolitan area. We talked about the plans for research, but as might be expected, we were asked many questions about indoor issues and cleanup crews. I talked about the recent announcement about lead. The audience was hungry for new information. The meeting did educate me about what the public was generally thinking and their worries about the future. Since I had the first opportunity to discuss the residential dust and lead issues, the public became much more aware of the need to properly clean up their homes. It was clear, however, that air-quality issues were getting mixed up with dust issues.

My primary point during the discussion was that although the lead levels in the ambient air were going down, the residences that were not clean could have lead loadings on the floor that were significant to health. These residences and other buildings needed to be cleaned up properly. At that time, some people were cleaning up residences on their own, others were having it done professionally, and others did not know what to do. During our individual talks, Lung Chi and I each made the point that if the levels on indoor surfaces were above a fraction of an inch, people should seriously consider professional cleaning. However, not being an expert, I felt that the local agencies needed to define a clear plan, which was still not available. It was a pity that some of the major issues discussed on October 18 were not widely reported in the press. This is one of the few times I would have been more than happy to sit down and discuss the issues at hand with any reporter. Those discussions, however, only began to occur later in the year.

That night, I retreated back to New Jersey, unsatisfied with my performance. I could not do much to allay the many fears. No one could.

From then on, I kept receiving phone calls about indoor issues and cleanup. In some cases, I explained to individuals that they should not be keeping pillows since they were full of dust and could not be easily cleaned. In other cases, I explained that rare and old books could be cleaned and preserved for future use. The days were intense, but I felt that we were trying to make a difference.

By the end of the month, I was close to having all of the data in hand from each investigator. However, there was still no word on permission to collect indoor samples, even though the need appeared to be getting more urgent. I saw the issues of asbestos resurface, which probably was reasonable for indoor cleanup and indoor ventilation systems, but the asbestos was not present as the traditional long thin fibers.

FURTHER THOUGHTS ON DUST

When I was a teenager driving dusty roads, the aerosol generated when my car tires would rub against the ground and release particles that were between 0.1μm to more than 40μm (10^{-6} meters) into the air was interesting to watch. These were primarily the inspirable particles defined earlier. That road dust material could stay in the air for a while. After about five minutes, the visibility in the air behind my car went back to normal, as if I had not been there. These materials in some ways mimicked the resuspended particulate matter that we defined as the WTC dust. As noted earlier, in the United States we were not necessarily concerned with these particles anymore because changes in industrial and energy activities had reduced the levels of very coarse particles and led to more emissions of the fine or respirable particles. We had been concerned about gas emissions that reacted with sunlight to form new particles from chemical reactions. These particles appeared to be especially toxic materials because they were readily available to the gas exchange region of the lung. This is probably why the EPA only analyzed its dust samples for particles that were less than 44 μm in diameter. Also, the very fine particles (less than 1.0 μm in diameter) stay in the air for a long period of time, which is in contrast to the short lifetime of inspirable particles in the air. In fact, if you leave suspended fine particles alone in a room, they will take a long time to deposit to a surface.

Remember, the collapse of the WTC towers generated a large and intense cloud of dust and smoke, and as Cliff Weisel quoted, it was like "a plague of darkness was upon the land that was tangible. It was so thick that no man could see his brother or rise from his place" (Exodus 10: 21–23, Hebrew numbering). Dust had been distributed everywhere in southern Manhattan. It was unfortunate that the particle size distribution was unknown because that was one of the first major pieces of information needed to start considering the potential sites of deposition in the lung and other parts of the respiratory system. It was estimated that together, the two WTC towers weighed ten million tons, including rock, mortar, cement, and other materials inside the buildings. Each building disintegrated to dust.

Since the initial dust particles clung to the clothing of survivors and was on the faces and in the lungs of the people who were caught in harm's way, there was the possibility of substantial one-time inhalation of the dust by commuters. They were carrying it home with them or carrying it around on their bodies through Lower Manhattan until they reached a place where they could change their clothes or could have their clothes washed down or decontaminated. The images were rather stark. Photographs of the potential acute exposures to people caught in the WTC dust clouds appeared on the pages of many magazines, including the cover of the October 8, 2001, issue of *Forbes* magazine. However, the most poignant is still the AP photo that was shown in this book as figure 5.1.

Over time, residents and owners who came back to their homes found the dust on their tables, chairs, furniture, toys, and clothing. Visible coatings of up to two inches thick were apparent. Without respiratory protection, these folks would inhale or ingest some of that dust. During the first month after 9/11, I received a number of calls from people in southern Manhattan who wanted to reenter or had reentered their residence and were concerned about the WTC dust. So, the dust was everywhere and exposures were possible.

I was at a meeting about three years after the collapse, and somebody asked, "How can you describe the dust?" Well, by that time I had the benefit of at least two years of analysis of samples by multiple scientists, friends, and colleagues and countless explanations of the WTC dust in the press. My answer to that question was actually simple: The dust contained "everything that we hold near and dear to us." Now there was a loaded answer that even evoked a follow-up from Anthony DePalma, then of the *New York Times*, and even another article about the dust in November 2005. I had not really thought of the implications of what I said, but to this day I think the answer was correct. The WTC dust held everything we consider near and dear to us. Obviously the answer contains reality and symbolism, the materials that composed the dust, the remains of souls lost in the attack, and the remains of two symbols of America's economy and place in the world.

8

WHAT WAS IN THAT WTC DUST?

What was in the dust that deposited all over southern New York City and on the many people who found themselves in harm's way, as captured in figure 5.1? First, there were no chemical, biological, or radiological weapons of mass destruction released by the terrorist act, and the material released was not similar to an industrial accident. To make sense out of it, we begin by considering it to be a combination of dust and smoke from the fires. However, the dust would also contain jet fuel that hadn't burned but could eventually burn after saturating the building materials that turned to dust. Therefore, what rained down during the rapid collapse was dust, smoke, and jet fuel.

Since the dust contained the material used to build the two towers, the dust contained disintegrated building materials. We validated the facts that building materials including cement, steel, other metals associated with steel and other construction materials (e.g., gypsum), and man-made fibers were in the dust. Because people worked in the offices and carried on their daily activities in the various businesses that were housed in the WTC or were visitors to the restaurants, stores, and observation deck since the mid-1970s, the WTC dust also contained cellulose from paper, napkins, and other paper products, but it was not present as free molecules since it was used in paper products. Furthermore, the WTC dust contained synthetic materials used in the fabrication of rugs, wood, and computer residuals. Human hair had been deposited on the rugs over the years. The presence of human hair might make you wonder why we did not measure it for DNA. The simple reason was that almost all of the human hair would have been accumulated over the many years that the towers were occupied. Therefore, these historical hair follicles would be an overwhelming contributor to the hair found in any sample, and would have yielded many false-positive values. For instance, I had been in the buildings a number of times, and my hair could have dropped on the floor of offices and hallways. The DNA of concern for the collapse would have been present with fresh blood and/or tissues.

The dust also contained a large amount of an unusual material: glass fibers. Both towers were 110 stories high and had 880 stories of windows in total since there were 110 stories with four walls times two sides of windows. These windows had been crushed by the collapse of the towers. These were different than other materials since the glass windows disintegrated into fibrous dust as well as glass chards.

One major component of the WTC dust was the dust particles that contained cement, which looked like grains of sand. Many of these grains could not be seen by the naked eye but were easily seen under the microscopes used by Jim Millette. Figure 8.1 depicts the component particles that were observed during the morphological analyses. The solid white line on the bottom of the figure shows the scale, which is 100 µm in length (one–ten thousandth of a meter). Thus, many of the particles were very long, and some were very large and had odd shapes (triangles and rectangles). Other particles were less than 5.0 µm in diameter. The point is that most particles appeared to be in the super-coarse particle size range.

The review of the analytical data provided by our colleagues found that over 50 percent of the mass of the WTC dust was made up of the cement and carboneous materials. (Carbon is the fourth most abundant element found in inorganic and organic forms.) As part of the carboneous materials, there were significant quantities of cellulose (paper). Most of the rest of the mass was made up of the glass fiber materials. Some of the glass fibers were formed from the disintegration of glass windows. Other fibers released during the collapse would have been part of interior wall board and ceiling tiles. Some of the fibers were characterized as slag wool and would become known as a specific and

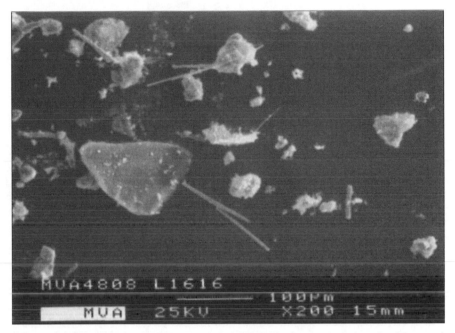

Figure 8.1. Detailed micrograph of the WTC dust sample taken by Lioy and Weisel. Taken by MVA Project 4808, MVA Scientific Consultants. Lioy et al. *EHP* 110 (2002): 703–714.

identifiable component of the WTC dust. Slag wool would become the second most famous material released from the WTC—not in terms of its health effects but in terms of a debate about whether it provided a usable tracer for identifying WTC dust indoors. All the materials that were fibrous but were not formed by the crushing of windows had been used to construct the interior walls and furnishings of the building.

ASBESTOS AS A COMPONENT OF THE DUST

Only one building had asbestos sprayed on approximately 40 percent of the beams used for insulation and fire protection; therefore, asbestos wasn't going to be the major part of the mass. Indeed, we found asbestos to be a minor part of the mass deposited on surfaces. However, because of the number of inches of dust we sampled on a surface, asbestos could be found in high loadings on surfaces inside buildings. Therefore, because of the large surface quantities and the potential for asbestos resuspension in a poorly cleaned or uncleaned residence, a professional cleanup would be required.

We found the asbestos concentrations in the WTC dust to be between 0.8 and 3 percent of the mass, but they were not of the typical sizes and shapes that we usually think about for serious asbestos exposure. Normally, the asbestos standard has focused on particles measured with Phase Contrast Microscopy that are longer than 5 µm and have a length-to-width ratio of 3:1. As I reviewed the micrographs taken by Jim Millette, it was clear that the asbestos was not released as a uniform fiber size or narrow size range, as would be the case during manufacturing, but there was a very wide variety of "structures." Thus, care had to be taken when interpreting the data presented about the total amount of asbestos that could be released into the air. That is also why many of the air samples were not above the standard of seventy fibers per cubic centimeter air. What it did indicate, and what was never truly relayed or understood by the public, was that although asbestos was released and there were asbestos structures in the range of concern for health, the amount of potential exposure was decreased because of the method of release of the asbestos. The wise suggestion by many scientists was that we have to wait and see, since it would take ten to twenty years to determine whether short-term exposure to high levels of asbestos will lead to chronic or fatal disease outbreaks, especially among those at Ground Zero. Clearly, I hope this does not come to pass.

PARTICLE SIZES OF THE WTC DUST

The issue of particle size was very important for interpreting the impact of the WTC dust and was a main concern of mine during the many discussions I had with the public, community, and professionals. As described earlier, the size of the dust particles was very important because it helps to define where the material will deposit in the lung and then provide clues about what would happen over time. Furthermore, it would help people understand the necessity for various cleanup strategies. Basically, the three size ranges measured in the samples were the fine, coarse (inhalable), and supercoarse (inspirable) particles.

The percent of the particles in the WTC dust were characterized according to size shown, and the percentages are in table 8.1 (for the details, see table A.1 in appendix A). To my surprise, greater than 98 percent of the mass was above 10 μm in diameter, which I defined as supercoarse or inspirable particles. This result was an important new piece of information about the WTC dust. The WTC dust was not distributed among the size ranges that one would typically find in the ambient air of New York City or any nondesert city in the United States. Modern-day particle air pollution in New York City is distributed primarily with the mass in the fine particle range, and with only 20 to 40 percent in the inhalable size range. In contrast, the WTC dust had less than 2 percent of its mass in fine and inhalable size ranges. Therefore, any comparisons of the initial WTC dust concentrations with typical NYC ambient air had to be made with caution because there would be little similarity between where WTC dust and typical NYC pollution would deposit in the lung. In addition, the concentrations of particles typically found in the ambient air in NYC during the early years of the twenty-first century were orders of magnitude lower than found throughout most of the twentieth century prior to the 1980s.

What was also of interest was the measurement of the aerodynamic diameter of the particles. This analysis, completed by Lung Chi, allowed us to estimate where in the respiratory system particles would deposit. Based on many years of research on lung deposition, including important experiments completed at the NYU Institute of Environmental Medicine, the large (supercoarse) particles would have a very difficult time penetrating deep in the lung. However, we must also note that no research had ever been done on the toxicology of such a mixture as WTC dust. The latter is understandable—who could have predicted such a complex material that was not normally found in the air, and in a very large particle range, would be formed or inhaled by the workers and residents of southern Manhattan?

Table 8.1. Outdoor WTC Dust Separated According to Particle Size, for Combined Samples from Lioy and Weisel, 2001

Size range (in micrometers)	0–2.5	2.5–10	10–53	>53
Percent of WTC dust mass	0.88–1.33	0.30–0.40	34.6–46.6	52.2–63.6

Since WTC dust was not in any way similar to typical modern air pollution, with much of the mass in or on particles greater than 10 μm in diameter, large particles associated with the entire range of the inspirable particles would behave similar to the road dust discussed earlier upon resuspension into the air by mechanical processes. Since about 2003, I have called these particles greater than 10 μm in diameter "supercoarse" particles. These are rarely seen in large quantities in modern American cities, but the term supercoarse provides a simple way to describe the WTC dust. The supercoarse particles emitted from the WTC would be inhaled as inspirable particles but would deposit mainly in the upper airways of the lung. This point would become important in describing the symptoms experienced by workers and others caught in the plume.

In addition, there were many fibers in the WTC dust. These fibers are cylinders that have a small "end-on" diameter and are long but not very wide. In some cases, they were 0.0001 to 0.001 meter in length, but still small in end-on diameter (less than 5 μm, or 0.000005 m). If such fibers got into the lung, they would be hard to remove. Think of it this way: you have logs going down a river along the stream lines, then they pass through a small opening into a pond. How easy would it be to get the logs out? It would be difficult because you have many directions to orient the logs and only one opening to return to the river. Thus, the probability of lining up to get back out is small. The phenomenon is similar for fibers. The fibers are inhaled and follow a path that moves air deeper into the lung. For those particles that do continue past the lung branches within what is called the bronchial tree to the alveoli in the gas exchange region of the lung (see figure 1.5), they accumulate in the alveoli sacs and stay there unless consumed by the macrophages (white blood cells). This phenomenon increases the likelihood of the fibrous particles finding locations to deposit in the lung both in the upper airways and the gas exchange region. In the end, this is probably what happened because of the large numbers of all particles that were in the air during the first few days postcollapse. This point is important since it would provide the opportunity for an individual to experience both short-term and long-term effects. Clearance of particles from the upper airways can take days, but clearance from the gas exchange region can take months or never happen.

DETAILS ON THE BASIC MORPHOLOGY

From the analytical results we received for the three dust samples that were taken at Cortlandt, Market, and Cherry Streets, the following basic characteristics were recorded (for more details, see table A.1):

Pinkish-gray color
pH: between 9 and 12
Nonfiber: between 30 and 50 percent
Fibrous: about 40 percent
Cellulose: between ~9 and 20 percent
Asbestos: between 0.8 and 3 percent
Trace quantities of many materials

We all knew that the appearance of the dust was very unusual and the characterization of a pinkish-gray color was a unique characteristic. Over time, the color changed to a light gray, suggesting that chemical or physical transformations had occurred. Furthermore, the WTC dust had a fluffiness or lightness in character and texture. This was probably due to the large amount of cellulose and fibrous material present. I had never encountered anything like it and will probably never again. To this day, I can pick out WTC dust—it is that unique.

The fibrous material was composed of disintegrated material that was present in building interiors or on the exterior. As time went on, we would call a large portion of it slag wool, based on work by the United States Geological Survey. However, there were other types of fibers that became of interest, including glass. The composition of the nonfibrous portion of the dust is simple to understand. It was the vast quantity of pulverized cement used to construct the building. The magnitude of the impact of the cement on human health would be found to be enhanced by the high pH of the samples, which indicated alkaline composition of the nonfibrous portion of dust. This variable did become important in the evaluations of exposure and health issues.

The pH is the measure of the acidity or alkalinity of a solution. It is a measure of the activity of dissolved hydrogen ions (H^+). In pure water at 25°C (75°F), the concentration of H^+ equals the concentration of hydroxide ions (OH^-). This will define a material as "neutral" and would correspond to a pH value of 7.0. Solutions in which the concentration of H^+ exceeds that of OH^- have a pH value lower than 7.0 and are known as *acids*. Solutions in which OH^- exceeds H^+ have a pH value greater than 7.0 are defined as *bases*. The higher or lower the number, the more pH can affect materials and humans. At a low pH, water will become corrosive (like sulfuric acid); at high pH, it will become alkaline (like Drano). The pH's of the three samples were alkaline and ranged in value from 9.0 to 11. The reason for the differences was the fact that the Cortlandt Street sample was collected in a protected area, and the other two were wet because of the rain on Friday the 14th. The rain caused some neutralization of the sample.

The observation of a high pH (above 7) in all samples, and the changes in pH in the total WTC dust samples that were subjected to rain, led to two important conclusions: First, the settled WTC dust could be easily neutralized when deposited from the ambient air and over time was subject to the weather and/or moisture. Second, the initial dust or dust that was left undisturbed in homes would be capable of irritating the upper airways of the lungs depending on the amount that was available and inhaled by a person without respiratory protection. The changes in pH also provided an indication of how the composition would change under various weather conditions, which again led to more complexity in understanding individual situations for exposure over time. For instance, for those people who came to Ground Zero during exposure period 1, the WTC dust was at its highest concentrations in the air, and the alkalinity was the greatest with a pH of about 11. The pH may have been higher or lower in other samples because of the collection location and the impact of rain and moisture. However, pH would continue to be an important indicator of significant exposures to workers and others.

During exposure period 2, the concentrations were lower, but the alkalinity was still high. After the rain events that occurred during exposure periods 3 and 4, the alkalinity of the WTC dust would go down outdoors, as seen in the Market Street sample. For these situations, the ability for any WTC dust resuspended outdoors to irritate the lung would also decrease. As a result, those working or arriving outdoors at Ground Zero after exposure period 2 would experience less in the way of exposures leading to health effects from the dust. For exposure period 5, the indoor situation, we still had no data. However, the pH results for the ambient samples would suggest pH values somewhere between 9 and 11, which should be a major concern for those who did not wear respiratory protection during the first few days post-9/11. The pH needed to be measured for the samples collected indoors.

The results also indicated that storage of the samples over a long period of time would influence the materials, variables, and the compounds that could remain stable or change form. The most stable materials would be the slag wool and other fibrous materials. Ionic and organic species would probably neutralize, transform to other compounds, or evaporate during storage.

ELEMENTS AND IONIC COMPOUNDS

The quantity of trace materials (which is really a poor choice of words from the standpoint of public health) describes the amount mass of specific materials that were present in the dust. To understand, however, the potential impact of the dust on health, it was necessary to know the levels of these trace materials. For example, lead was only present in WTC dust in trace amounts, much less than 1 percent of the mass. It is a highly toxic metal, however, and needed to be of concern at the levels encountered on outdoor and indoor surfaces around Ground Zero. So, one of our goals was to determine how many other toxic materials were in the dust, and whether the levels posed a hazard if there were short-term high exposures or long-term exposures. Therefore, we measured trace materials that were present in the WTC dust sample.

The trace quantities of twenty-five elements measured in the WTC dust, including the lead mentioned earlier, are presented in detail for the three dust samples in tables A.1 and A.2 of appendix A. These elements were detected in all three samples, and the range of concentrations found in the samples are reported at relatively low levels, in the part per billion range (which means one molecule of an element for every billion molecules). The majority of the elements detected could be assigned to building materials as the source. These elements included magnesium, aluminum, barium, chromium, and manganese. The two elements with the highest concentrations were titanium and zinc, in the thousands of part per billion range (called parts per million parts). These would be

found in white paint and building materials, respectively, and were identified in some of Jim Millette's micrographs and on the surface of the
wallboard sample that is still in my office. Further, having entered the
towers over the years I had a firsthand look at the white paint.

The presence of lead has already been discussed, and the source was
probably lead-based paint that was still allowed to be used on metallic
surfaces at the time of the construction of the WTC.

In contrast to the above, and as a bit of good news, the toxicants mercury, arsenic, and cadmium were present at very low to nondetectable
levels in the dust. In the early days post-9/11, people were expressing
concerns about the presence of mercury. The source would be the
release of mercury from lightbulbs during the fires, which would also
reduce the levels that remained on the dust because of the high probability of evaporation during the fires. The previously mentioned biological monitoring completed on firefighters by David Prezant found no
accumulated mercury levels in biological specimens taken from these
individuals. Thus, the WTC dust had low levels of mercury, and the
firefighters had low levels of mercury. The point made by many was that
mercury would have been present in the lightbulbs, but that contact
would have been difficult given the way in which the source would have
released the mercury. The volatile nature of the material would have
led to its release when the lightbulbs disintegrated or were burned in
the fires. So, yes, there was a hazard, but no, there was little contact and
exposure. Therefore, the risk to workers would have been low.

The ionic species found in the dust were in concentrations of mil-
ligrams of ion per gram of total dust. There was a misprint in the original paper, which reported the levels of ionic species to be a factor of
1,000 lower than actually reported. However, this did not change our
interpretation of the results. The units have been corrected in appendix
A. The ionic species again fit with the composition of the dust being
dominated by construction material debris. Calcium and sulfate were
found at the highest concentrations, having a range of 14,000 to 18,000
parts per billion and 35,000 to 45,000 parts per billion respectively, and
each would be a marker for a very common building material: gypsum
($CaSO_4$-$2H_2O$), which is a naturally occurring crystal. There was some
attempt to link the sulfate to the fires, but this would be minimal in
comparison to the amount of gypsum that would have been present in

the interior walls of the towers. As time went on, I was given artifacts from the WTC buildings. One was a two foot by two foot piece of wallboard, which would contain gypsum. It sits next to my WTC gear in the office. A colleague at the medical school just handed it to me one day. Other ions detected in the WTC dust in just the part per billion range were chlorides and sodium.

The microscopic analysis of three samples of settled dust brought many of the semiquantitative and qualitative results together. I offer the following summary of all of these results, which is paraphrased, with permission from the publisher, and expanded on from the scientific paper published in *Environmental Health Perspectives* by Lioy et al. in July 2002 (see the bibliography for reference).

CORTLANDT STREET WTC DUST SAMPLE

The Cortlandt Street sample was collected about two hundred to three hundred feet from Ground Zero in an area sheltered from the weather. The WTC dust contained construction debris, and the materials detected included vermiculite (a natural insulation material), plaster, synthetic foam, glass fragments, paint particles, glass fibers, lead, calcite grains, paper fragments, quartz grains, low-temperature combustion material (including charred woody fragments), and glass shards. Chrysotile asbestos fibers were present and were estimated to be less than 1 percent of the sample by volume, and much of the chrysotile was stuck to the carbonate binder material that attached it to the building. Thus, these were not free asbestos fibers; they were a postapplication form that had been dislodged from the beams of one WTC tower. Some skin cells and dyed cotton were also detected in the WTC dust. The finding of skin cells was consistent with the types of skin cell usually found in dust in any indoor environment, especially in carpets. They were not fresh remains from those who had died during the collapse.

Approximately 35 percent of the volume of this sample was in the form of loosely attached clumps of fibrous material and included primarily glass fibers, and slag wool. An example of the typical glass fiber is shown in figure 8.2. In many cases, the width was $\cong 1$ µm (to greater than 10 µm), and the length ranged from 5 to 100 µm. The fiber shown in figure 8.2 is not a "clean" glass fiber; there were other materials attached along the rod. The fiber was around 1 µm in width which means that it could easily deposit in the bronchial airways and maybe even reach the gas exchange region of the lung! The attachment of other materials to a rodlike particle was a typical feature noted for many different fibrous particles identified in each sample. Many had significant quantities of the attached material. We speculated that, if inhaled by a person, the fiber itself could easily lodge in an upper airway of the lung, and the loosely bound material of smaller size could have been released over time and probably reach the gas exchange region of the lung. Then the fibers would have deposited in the alveoli. This was an idea, not necessarily a reality, but it was a reasonably educated guess.

The scanning electron microscopic analysis of the fraction of particles less than 75 µm in diameter still found most particles to be in the size range that included inspirable material—the supercoarse particles—and revealed many glass fibers and cement particles. Some fibers contained calcium, silicon, and sulfur. Some particles were composed of calcium carbonate (figure 8.3).

Chrysotile asbestos fibers, identified by transmission electron microscopy, were found in the size range that was less than 75 µm in diameter. None of the analyzed particles contained lead, chromium, cadmium, or mercury, although lead chromium and cadmium were found in this sample by elemental analyses. The reason for the difference is due to the small number of particles that can be reasonably subjected to detailed microscopic analysis and the trace quantities of the material present in the samples.

CHERRY AVENUE WTC DUST SAMPLE

Similar to the samples collected by the USGS, the Cherry Avenue sample was not collected in a covered area; however, it was a very large sample because it was taken from a location that had over six inches of WTC dust on its surface. Similar to the Cortlandt Street sample, it was mainly composed of construction debris, including cement, vermiculite, plaster, synthetic foam, glass fragments, mineral wool fibers, paint particles, glass fibers, metals, calcite grains, paper frag-

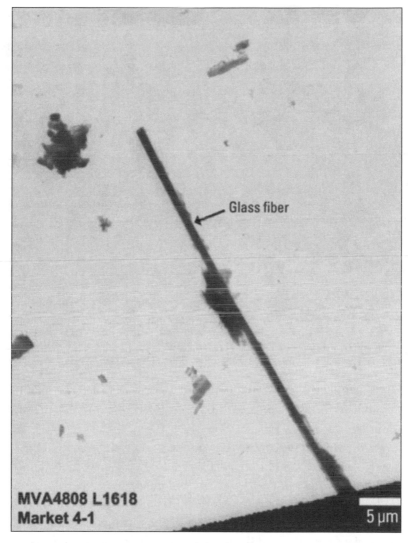

Glass fiber

MVA4808 L1618
Market 4-1

5 µm

Figure 8.2. A glass fiber in the WTC dust with many other smaller-sized particles agglomerated on the surface. Electron micrograph taken for MVA Project 4808, MVA Scientific Consultants. Lioy et al. *EHP* 110 (2002): 703–714.

ments, and quartz grains. Again, there was low-temperature combustion material in the WTC dust that included charred woody fragments from the fires and metal flakes. We estimated that in this sample, the chrysotile asbestos fibers were less than 1 percent of the sample by volume, and consistent with the above much of the chrysotile asbestos

Cortlandt 2-3   50 µm

Figure 8.3. A supercoarse calcium carbonate particle in the WTC dust sample collected by Drs. Lioy and Weisel. Electron micrograph taken for MVA Project 4808, MVA Scientific Consultants. Lioy et al. *EHP* 110 (2002): 703–714.

had the carbonate binder attached to it. We observed some hair fibers that would be found in rugs, and tarry fragments were also present in the sample.

Approximately 10 percent of the volume of the sample was in the form of loosely consolidated clumps of fibrous material, of which the greatest portion was glass fibers. The scanning electron microscope analysis of the fraction less than 75 µm in diameter again revealed many glass fibers and cement particles, some in a fibrous form, containing calcium, silicon, and sulfur. We used scanning electron microscopy and transmission electron microscopy (TEM) to examine chrysotile asbestos fibers, lead paint fragments, iron-chromium particles, and soot particles found in the less than 75 µm fraction. The soot particles were in the submicron size range, and a vivid example is shown in figure 8.4. No particles containing cadmium (although detected by ICP/MS) or mercury were found above the minimum detection limit in the one thousand particles analyzed from this sample. Particles smaller than 75 µm in diameter included asbestos, soot, and lead.

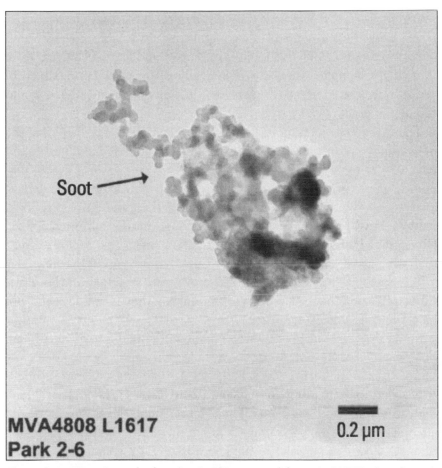

Soot ⟶

MVA4808 L1617
Park 2-6
0.2 μm

Figure 8.4. Soot from the burning building materials at the WTC after the collapse. WTC dust sample collected by Drs. Lioy and Weisel. Electron micrograph taken for MVA Project 4808, MVA Scientific Consultants. Lioy et al. *EHP* 110 (2002): 703–714.

MARKET STREET WTC DUST SAMPLE

The Market Street sample was the largest sample, giving us well over a pound of WTC dust that we collected on September 15. Portions of this sample have been sent to a number of investigators during the intervening years for detailed analysis. At this time, however, I see this sample as having only historical value since the sample is over seven years old. Over that length of time in storage, all reactive species would have changed form, semivolatile compounds and volatile species would have evaporated, and the ionic structure could have changed, even though the samples were stored in a cold room at a temperature of about 37°F. The nonreactive materials, like elements and fibers, would remain intact.

The Market Street sample also had construction debris, including the now familiar vermiculite, plaster, synthetic foam, glass fragments, paint particles, mineral wool fibers, glass fibers, calcite grains, paper fragments, quartz grains, low-temperature combustion material (including charred woody fragments), and metal flakes. Chrysotile asbestos fibers made up less than 1 percent of the sample by volume, and much of the chrysotile adhered to carbonate binder. This result was different from the bulk mass results, which indicated that 3.0 percent of the mass was asbestos. A difference of a factor of three times higher can be easily expected, since the sample was not homogeneous, meaning it did not have the consistency of a puree.

Some dyed cotton fibers, tarry fragments, pollen grains, and metal flakes were also present. Approximately 10 percent of the volume of the sample was in the form of loosely consolidated clumps of fibrous material, called, in general, man-made vitreous fibers (MMVFs), of which the greatest portion was glass fibers. The scanning electron microscope analysis of the fraction less than 75 µm in diameter revealed many glass fibers and cement particles, some in a fibrous form containing calcium, silicon, and sulfur.

Chrysotile asbestos fibers, identified by TEM, were found in the fraction less than 75 µm in diameter. We found no particles containing lead, chromium, cadmium, or mercury in the single particles analyzed from the Market Street sample, although all but mercury were detected by the inductively coupled plasma/mass spectrometry analyses of the bulk sample reported earlier.

As stated previously, the morphologic differences among each of the collected samples were minor, and the differences noted could be attributed to the fact that we analyzed only one thousand particles per sample. The limitations in the number of particles that can be reasonably analyzed in detail would preclude consistent detection of the presence of materials that comprise less than 0.1 percent of each sample.

This was sort of a roll of the dice for material present in very small quantities. Fortunately, the other analytical methods filled in many gaps concerning the overall composition of the WTC dust.

In summary, the morphological analyses completed on the three samples showed there were only minor mass differences between the Cortlandt Street and the Market Street and Cherry Street samples. For example, the Cortlandt Street sample had 0.88 percent fine particles (particles less than 2.5 μm in diameter), while the other two samples had greater than 1.1 percent in the fine-particle mass. Using microscopic analysis to generally describe the distribution of materials among the mass fractions, we found that large, supercoarse particles were primarily made up of building materials including gypsum, glass fiber, MMVF, wood fibers, and paper fragments. Chrysotile, which adhered to building material, chrysotile bundles, and plaster, was also a component of large particles. This is consistent with the fact that the lint with fibrous particle bundles was also in the greater than 300-μm particle size range. The glass fibers were also very curious. Since the buildings were constructed with 110 stories of glass windows, many windows were crushed during the collapse. They, too, had returned to dust, resulting in many small glass fibers. These fibers could be highly irritating upon inhalation into the lung and deposition within the upper airways.

After a presentation of our results to the Acute Exposure Guidelines Committee of the National Research Council early in the summer of 2002, I received a piece of ceiling tile, circa 1973, which would have been used in commercial installations like the WTC. The disintegration of the many square feet of this material, or similar material would have also contributed to the glass fibers present in the debris. The MMVF has been reviewed by the National Research Council, and although MMVF has been found to have a much reduced level of toxicity compared to asbestos, it still is a lung irritant. Therefore, because such high levels were found in the WTC dust, MMVF became a chemical of concern for residential cleanup.

The particles less than 75 μm in diameter were still primarily supercoarse particles, and these included asbestos, soot, lead, and other trace elements. This is consistent with the dual nature of the event—the collapse of two buildings overlaying intense and uncontrolled burning structures—which would result primarily in individual and population exposures to large particles and in much lower exposures to fine particles. However, the large amounts of material in the air during exposure periods 1 and 2 postcollapse would still lead to short-term high exposures to fine particles as well as the inspirable particles among unprotected individuals.

In addition to cellulose in paper products, the WTC dust contained large quantities of organic carbon material that were the by-products of combustion caused by the attack and subsequent fires. The combustion products contained materials that were semivolatile (could be found on a surface or in the air depending on the temperature, which also means they may not have completely evaporated; the semivolatile compounds would stay in the dust as long as it wasn't too warm). These organic compounds would come from the composition of the unburned jet fuel that dripped onto the buildings and then remained on the surface of the dust particles after settling, or portions of the mass that burned. As a result, the dust contained polycyclic aromatic hydrocarbons, which are carcinogens but are released by all kinds of combustion that occur every day, including fires and industrial production combustion events. Some of the other materials burned during the WTC fires were plastics and other synthetic consumer or office products. Thus, the particles would also contain phthalates, and possibly some dioxins, in addition to the previously mentioned unmeasured combustion gases.

ORGANIC MATERIALS IN THE WTC DUST

This discussion will be organized into three sections based on the type of organic compounds: polycyclic aromatic hydrocarbons, semivolatile species, and other organics. Tables with representative examples of the analytical results for the organic materials are found in appendix A.

Polycyclic Aromatic Hydrocarbons (PAHs)

These are a class of chemicals routinely found in oil and coal, or they can be formed during the burning of materials, either within an uncontrolled fire or within a controlled combustion process. They consist of fused aromatic rings. The PAHs are of interest to the environmental health science community because some compounds have been identified as carcinogenic (causing cancer), mutagenic (causing mutations of genes), or teratagenic (causing reproductive effects). The names and the levels of a suite of PAHs that we measured in the WTC dust are found in table A.3, in appendix A. Relatively low concentrations were measured for all compounds, in the range of parts per billion, which means one molecule of a PAH for every billion molecules of dust. The results did indicate that the initial WTC dust did have PAHs from the burning fires.

To put this finding in perspective, these concentrations were overwhelmed by the mass of cement and other building products. One scientific paper by Offenberg (reference provided in the bibliography) provided the results of a detailed analysis of the PAHs. For a summation of thirty-seven individual compounds, these researchers found that PAHs made up about 0.04 percent of the mass. We also measured the same PAHs in three other samples from Lung Chi Chen, and the levels were slightly lower. One curiosity was that in contrast to typical combustion plumes, the highest concentrations of the PAHs were found in the larger, supercoarse particles and not only in the fine particles. The fine-particle size range is typically where PAHs are found, released by well-controlled combustion processes like a power plant. Thus, the PAHs measured in the WTC dust samples would not be preferentially deposited in the gas exchange region of the lung. They would deposit in the upper airway, a place where it is easier to clear them from the lung.

Semivolatile Compounds

For the WTC dust samples, the main contributions to this general class of compounds were produced during incomplete combustion of the jet fuel in the fires, and chemicals that were bound to or part of the droplets of unburned jet fuel. There were approximately ninety-one thousand liters of unburned or burning jet fuel, and unknown amounts of fuel stored below the WTC towers. A significant product of incomplete combustion found in all three samples was the class of contaminants called phthalate esters (shown in appendix A). The levels of phthalate esters were greater than 10 µg/g for specific compounds. The total level of phthalate esters in the Market Street sample was greater than 100 µg/g.

The variety of hydrocarbons measured by the GC/MS analysis were identified as a result of comparisons with known peaks that provide a fingerprint of each compound. Some of these results are shown in figure 8.5a and 8.5b. The analysis showed that jet fuel was present because in each sample we found an envelope (a smooth rise below the individual peaks of high boiling hydrocarbons of ten carbons or greater), as well as individual compound peaks on the envelope. These figures also tell so much about how the detected hydrocarbons were consistent with saturated hydrocarbon chains and naphthalene ring structure, and the data were very helpful in source identification. All samples also showed major peaks of the lightest PAHs (naphthalene, substituted naphthalene, acenaphthalene, and fluorene), which were consistent with the presence of products of combustion and supported the quantitative results reported in table A.4.

Hydrocarbon peaks were found by the GC/MS analyses in all the samples including those collected farthest to the east of Ground Zero (0.7 km; Market Street). These peaks are shown in figure 8.5a and

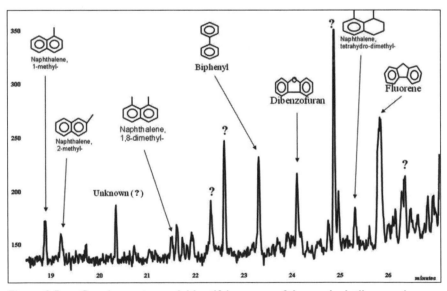

Figure 8.5a.   Gas chromatograph identifying some of the semivolatile organic compounds detected in WTC dust samples collected by Drs. Lioy and Weisel. Analyses by Brian Buckley's analytical core in coordination with P. J. Lioy at EOHSI.

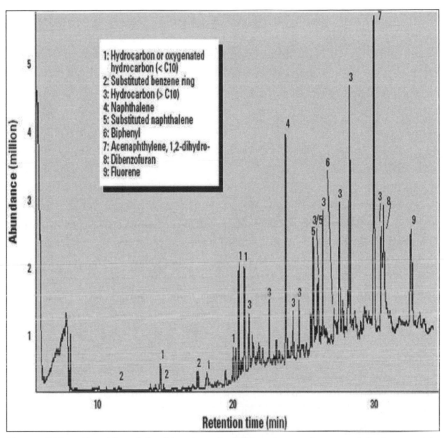

Figure 8.5b. **Graphic of the identity of additional organic materials detected in WTC dust samples collected by Lioy and Weisel. Analyses by Brian Buckley in coordination with P. J. Lioy at EOHSI. Lioy et al.** *EHP* **110 (2002): 703–714.**

8.5b. The alkane hydrocarbons detected were the same as those that would be found in uncombusted fuel. Jet fuel is composed of a mixture of saturated hydrocarbons (greater than 50 percent of the total fuel) and aromatic hydrocarbons. The most common types of compounds present were paraffins and cycloparaffins, which are common to kerosene and are found as part of jet fuel. We also identified some complex peaks of PAHs beyond the thirty-seven mentioned in the work done by Offenberg. Again these were a result of the burning of material from the buildings and the jet fuel. From all these data, the most interesting point was that the jet fuel was on the dust, and some of the material remained attached as a surface coating on the crushed building materials.

Particles as well as the unidentified and unquantified gases transported away from Ground Zero during the initial hours after the attack contained a mixture of combustion products and jet fuel. Thus, residents downwind in Brooklyn where the plume, even though diluted and well aboveground in most locations, touched down would have been exposed to particles from construction debris and products of incomplete combustion. Some of those particles were coated with jet fuel. Considering the amount of fuel that was released, even if other petroleum-based products were released from storage in the towers and the surrounding area, the levels would have been small in comparison to the amount released as jet fuel or the products formed during its combustion.

OTHER ORGANIC CHEMICALS

Other organic chemicals were detected including a number of classic environmental organic compounds. These analyses were done to help identify materials that could have long-term health effects besides the previously mentioned asbestos. The two most common are polychlorinated biphenols and dioxins/furans.

Polychlorinated biphenyls (PCBs) were used as coolants and lubricants in transformers, capacitors, and other electrical equipment because they don't burn easily and are good insulators. Products made before 1977 that contain PCBs include electrical devices with PCB capacitors. Manufacturing ended in the United States in 1977 because PCBs could cause human health effects and can build up in the environment. PCBs are associated with cancer of the liver and biliary tract. Pregnant women exposed to relatively high levels of PCBs have babies who weigh less than normal. The Department of Health and Human Services (HHS) has concluded that PCBs may "reasonably be anticipated" to be agents that cause cancer. The EPA and the International Agency for Research on Cancer have also determined that PCBs are probably carcinogenic to humans.

Dioxins and furans are a group of chemicals with similar chemical structure and a common mechanism of toxic action. Included are seven polychlorinated dibenzo dioxins (PCDDs) and ten of the polychlorinated dibenzo furans (PCDFs). Both the PCDDs and PCDFs are not found in commercial products but are by-products of combustion and industrial processes. Because dioxins are distributed widely in the environment, they are defined as persistent and can accumulate in the body. Most people have low levels in their body, only parts per trillion, and these will remain for years. EPA classifies dioxins as likely human carcinogens, and they can increase the risk of cancer. In 1997, the International Agency for Research on Cancer classified 2,3,7,8, TCDD, the most studied dioxin and a known human carcinogen. In addition, levels only ten times background levels, noncancer health effects have been observed from TCDD exposure in animals and, to a more limited extent, in humans. In humans these effects include changes in early development and hormone levels.

The polybrominated diphenylethers (PDBEs) and phthalates are less understood in terms of human health effects. The PDBEs are well-known flame retardants and are in the same class of chemicals that include PCBs. They have been utilized in building materials, electronics, furnishings, motor vehicles, plastics, polyurethane foams, and textiles. The three main types of PBDEs are penta, octa, and deca. The European Union (EU) banned the use of penta and octa-BDE in 2004 and deca-BDE in 2008. PBDEs have also been found at high levels in indoor dust, sewage sludge, and wastewater treatment plant effluents, and they have been found in breast milk. Concerns have been raised about the safety of PBDEs substances, such as penta-BDE which accumulate in human breast milk and fat tissues. According to EPA's Integrated Risk Information System, these chemicals may cause liver toxicity, thyroid toxicity, and neurodevelopmental toxicity.

The phthalates are esters of phthalic acid and mainly used as plasticizers (substances added to plastics to increase their flexibility). They are chiefly used to turn polyvinyl chloride (PVC) from a hard plastic into a flexible plastic. Heating or burning of PVC materials releases phthalates and other combustion products into indoor and ambient air. The PVC or vinyl phthalates are commonly used in many consumer products as a plasticizer or softener. Phthalates are also used in cosmetics and other personal care products, pesticides, building maintenance products, lubricants, adhesives, film, and medical devices. Phthalates have been identified as a potential endocrine disruptor, which has been demonstrated in some studies to affect male reproductive organ development. These have as a metabolite a monoester, which is toxic. Most concerns have been directed toward pregnancies, but recently the United States has banned the use of specific phthalates in children's toys, a primary route of ingestion exposure for toddlers.

The levels of PCBs and polychlorinated dibenzo-p-diozins (PCDDs) and dibenzofurans (PCDFs) found in the WTC dust were in the nanograms per gram and picograms per gram range (thus present at very trace concentrations), as shown in table A.5. Thus, the WTC dust yielded detectable, but not excessive, levels of these categories of environmental contaminants. The PCDD and PCDF dust sampled by EPA directly from the pile had maximum total equivalents of about 300 ng/kg. Neither our study nor the EPA found PCDD levels in dust above background levels.

The levels of polybrominated biphenyls and brominated diphenyl ethers (BDEs) are found in table A.6. The levels have been described as similar to those found in sewage sludge. The penta-mixture (BDE-47, BDE-99, and BDE-100) detected in the WTC dust was used in flame retardants for polyurethane foam, which is common as padding in office furniture. The highest BDE concentration was for BDE-209, which is used as part of thermoplastics in computers. The large volume of material present in the towers led to significant ambient levels of polybrominated biphenyls, BDEs, during the first day after the attack on the WTC. The pesticide chlordane was not found at levels above those that would be found as a background in soil.

The levels of specific phthalates ranged from nondetectable values to a maximum of about eighty parts per million, which were at or below the concentrations found in house dust. However, as with lead, these levels all needed to be considered in context. The total amount of dust in a residence must be taken into account. Even though the levels per gram of dust were relatively low, the total amount of dust present in a residence could lead to higher surface loadings. As I said in October 2001 and for many months thereafter, the bottom line was that residences and other buildings with more than a light surface loading would need a professional cleanup.

If one could measure to even lower concentrations using a more sensitive technique, WTC dust may have contained enough elements and materials to fill out a large portion of the periodic table. Thus, although we may not have detected all materials or all elements, I would assume that if one took a very hard look at the dust sample, more compounds would be found than what we did report. Again, the dust contained materials that were used to make the things that represent the raw material and by-products that defined our civilization. (Ah, this may be an interesting doctoral research project for a bright student in the future.) In the end, however, these results would have little impact on our understanding of observed health effects.

If the events of that day had quickly resolved themselves or if multiple days of significant rain had occurred immediately after the attack, we probably would have looked at the WTC dust more like a curiosity, sort of like how we reacted to moon dust. However, as we know, many people inhaled this dust in those first twenty-four hours postcollapse—and inhaled it in very large quantities due to the total magnitude and complexity of the dust released by the collapse and the disintegration of the buildings. The WTC dust, as we now know it, began to take on a life of its own since it wasn't a very healthy situation near Ground Zero for the emergency personnel in New York who did not use respiratory protection. This situation led to the involvement of more health care professionals who had to deal with the growing number of symptoms and medical problems among the workers.

For the scientists analyzing the dust, it seemed as though everyone was always looking for more and more detail. Unfortunately, the farther away analyses were from the event, the harder it was to identify and characterize apparent dust samples. I will discuss this later on in terms of the WTC signature, which become an object of debate rather than a means to an end during 2004 and 2005. However, the questions about the indoor environmental quality needed firm answers, and at that time there still weren't any that were readily available.

9

INTO THE UNKNOWN: EXPOSURE AND HEALTH

For the physicians in the NIEHS Centers and NYC departments, the WTC dust health concerns started during the second week post-9/11 when health professionals began to see firefighters and police officers coming down with a cough they had not anticipated; it became known as the WTC cough. This cough became a more public issue in the early part of October 2001. Based on some superb work by my colleague Dr. David Prezant, the WTC cough was found to be associated with the time workers arrived at the WTC and whether they had used respirators. The WTC cough appeared to be partly associated with the WTC dust, especially the dust that was present and then resuspended within the first twelve to seventy-two hours postcollapse (exposure periods 1 and 2). During this time, the fires continued to burn intensely, which meant not only would the dust be released, but also gaseous and particulate products of incomplete combustion would be emitted by the fires.

From the standpoint of exposure, we began to understand some material in the dust could cause a variety of acute effects. Those materials were not traditionally found in the air because they were associated with cement, which we normally use in construction and in solid forms on sidewalks and buildings. The cement blocks that were deconstructed into dust particles during the collapse were highly alkaline in nature and could irritate the lung. The pH measurement provided information on the alkalinity of the particles. Remember that for the samples that we and others took, the pH was above 9 and as high as 11. An educated guess, a hypothesis, would suggest that during the exposure periods 1 and 2, individuals inhaling these gases and particles as an aerosol could have had synergistic or additive effects caused by the aerosol inhaled. The evidence seemed to be pointing toward the alkalinity as one major reason for the health effects observed among responders.

THE UNKNOWN GASES IN THE WTC AEROSOL

I must temper this conclusion a bit. Although I have primarily focused on the WTC dust, I stated early on that it truly did not represent the whole story in terms of the magnitude of the situation encountered during exposure periods 1 and 2. When I first started this story, I was talking about aerosols and their characteristics. These aerosols are particles suspended in gases. The dominant gases that were present after the WTC collapse were still the "air," which contains primarily nitrogen (78 percent) and oxygen (21 percent). Prior to the collapse on September 11, the air was fairly clean because of local weather conditions. The 11th started as a nice clear day with brisk northwest winds blowing across Manhattan.

Unfortunately, during and immediately after the attack and collapse of the towers, gases were released by the fires that continued while the dust was spreading throughout southern Manhattan. So the "rest of the story" during exposure periods 1 and 2 was the large amount of unknown gases being released because of the combustion of materials (e.g., jet fuel and combustible products) that were part of furnishings and interior building materials.

In retrospect, during the first twenty-four to forty-eight hours post-collapse, we should have been calling the WTC dust the WTC aerosol. That WTC aerosol contained both the dust and the gases that were the net result of the emissions from collapse of the WTC. Unfortunately, we will never know completely what was in this WTC aerosol because we did not measure the gaseous phase of the combusted material released into the atmosphere. Thus, we could not define the magnitude, number, and types of gases that were released from the materials used to furnish and build the WTC that burn and release gaseous compounds. One can only surmise, but anything that was plastic or synthetic in nature would go up as gases if it caught on fire. Those gases have been known for years as being very toxic in human lungs. The combustion gases could contain benzene, aldehydes. toluene, cyanides, phosgene, nitro-PAHs, and more. If you put the WTC dust into the context of an aerosol, it is a part of the WTC aerosol but not the only part. The gaseous component was present simultaneously and would have included trace quantities of gaseous toxicants. However, we will never know the nature of that toxicity of the gaseous mixture and the resultant exposures because we could not take any measurements, or reproduce (simulate) the exposure in a toxicology laboratory during inhalation studies of laboratory rats or mice.

The EPA did studies on the fine-particle fraction of the dust, which showed minor irritations in the respiratory system of rats. Again, EPA worked on the size fraction that it regulates, not the total dust. The NIEHS did studies of total WTC dust, though never published them, that showed irritation in the lungs of rats for about a week after instal lation in the respiratory system, but no long-term effects. An interesting postscript: the NIEHS paper was reviewed, and the reason given for not publishing was that it was not mechanistic toxicological research! It was, however, research that was needed to help understand the reasons for the health effects being experienced by actual survivors. Still, obvi- ously something was missing since the animal studies could not account for all of the observed human health effects. My suspicion and that of David Prezant was that the variable missing from all of the toxicological studies was the magnitude and types of gases inhaled during exposure periods 1 and 2.

In 2003, I started to write a letter to the editor on this topic but never submitted it. I offer it here as a possible explanation about exposure to WTC aerosol:

Prezant et al. (2002) reported the incidence of cough and bronchial re- sponsiveness in firefighters who were at the WTC site during the rescue operations and/or beyond (e.g., more than seven days). In their analyses, it was clear that the closer the interval of time spent by an individual at Ground Zero was to 9/11, the higher the incidence of upper respiratory effects among firefighters. Unfortunately, the exposure data available for use by Prezant were limited because (1) no monitoring occurred in the first few days, (2) samples of dust and smoke were still being examined for composition, and (3) many of the early samples collected by either the state or federal government were examined for traditional compounds of concern. Lioy et al. (2002) published an article providing a comprehensive analysis of the dust and smoke that was initially released by the fire and collapse of each tower, and then settled as resuspendable dust and smoke throughout Lower Manhattan on 9/11. The approach used by Lioy et al. was to quantify as many of the physical and chemical properties of the settled dust and smoke as possible. This was based on the hypothesis that the dust generated by the collapse, and smoke caused by the burning jet fuel and other materials, would be a complex mixture composed of pulverized building material and combustion products. Included in the analytical scheme were organic and inorganic constituents, particle size distribution and morphology, and the bulk components of the mass.

Without knowledge of the Prezant et al. work, Lioy et al. hypothesized that the initial high exposures on 9/11 and the lower exposures a few days thereafter could have resulted in short-term health outcomes. Because of the timing of the events and general concern in NYC, these were initially noted in numerous popular press articles.

Subsequently, a number of the materials of concern for short-term and long-term health effects were reported by Lioy et al. and were included in the EPA cleanup program as targets for removal from indoor residences. The analyses clearly identified "nontraditional" or atypical air pollutants as materials of concern for all individuals who may have been in contact with the dust and smoke; included were glass fibers, and high pH (>9) cement dust particles. Other materials included were asbestos, lead, polycyclic aromatic hydrocarbons, and dioxins. For many of the particles present in the dust and smoke, the particle sizes were quite different than those routinely encountered in the ambient air. Furthermore, the short-term exposures to some of these materials could be a cause of related health effects.

As cited in the article by Prezant et al., most particles measured initially by environmental and occupational agencies were primarily asbestos. However, the work of Lioy et al. found that greater than 99 percent of the dust and smoke particles generated by the disintegration of the WTC were >10 μm in diameter; thus, the initial results on asbestos did not describe size ranges and mass contributions of concern for acute effects. In fact, between 37 and 74 percent of the particles were in the range from 10 to 53 μm in diameter, with the highest percentages associated with the indoor settled dust and smoke samples. The glass fibers, which comprised approximately 40 percent of the mass, were between 0.5 and 5 µm in diameter, and anywhere from tens to hundreds of micrometers in length. In addition, the glass fibers had many smaller particles attached to the surface of the rodlike structure. For the cement and other building materials, the particles were also primarily >10 µm in diameter and comprised over 40 percent of the bulk mass. Organics from combustion and cellulose comprised another 10 percent. Finally, the WTC dust and smoke was very light in density and easily resuspended in the air.

Each of these pieces of information about the settled dust and smoke indicate that most of the particles were in the inspirable particle size range (<100 µm in diameter). Thus, with respect to deposition within the respiratory system, the WTC dust and smoke particles would preferentially deposit in the upper airways, including the nasal-pharynx, trachea, and upper sections of the bronchi. The particle deposition processes of interception and impaction at various sites in the upper airways would have led to the accumulation of high-pH cement particles, glass fiber, and the scouring of smaller particles into the lower airways. These processes would have led to the upper airway deposition because of the unprecedented high levels of pH inspirable particulate matter loadings in the atmosphere during and immediately after the collapse. The exposure would have continued on subsequent days, although at lower levels, due to local resuspension of the particles after the collapse. Videos and still pictures clearly document the high resuspension exposure from outdoor and other surfaces. Each of these points would fit into the Prezant et al. empirical model for acute health outcomes. Furthermore, the fact that the WTC dust/smoke was a complex mixture of agglomerated particles leads to the hypothesis they would have deposited in the upper airways of lung throughout the first few days and hours post-9/11. Furthermore, the smaller agglomerated particles could easily slough off dropped fibers and move deeper into the lung. *However, the unknown and potential synergistic component of the WTC dust and smoke were the gases released at the same time as the dust. These were not measured and could have seriously affected the total inhalation burden among the firefighters and others who were not wearing respirators.*

Lioy has examined the potential for different exposure patterns that occurred post-9/11. In fact, there were five periods of WTC exposure, three of which were directly applicable to the initial dust and smoke (*see table 1 in this book*). The levels in exposure period 1 contacts would be much greater than 5 mg/m^3 (estimated) and would be characterized by the complex mixture of the alkaline dust and smoke and gases. Period 2 would be much lower concentrations and exposure to dust and smoke, but not insignificant (0.5 to 1 mg/m^3), caused by the reentrainment of the settled dust and smoke generated during the first days after the collapse. The levels of the gases were still basically unknown. Period 3 would have led to lower levels of dust and generated smoke since much of the outdoor dust was removed during two separate rain events on September 14 and 24, and the fires were primarily smoldering.

Particles associated with four of the five categories post–WTC attack could have affected the firefighters studied by Prezant et al. A fifth period, indoor environment, is not directly applicable to firefighters but is a concern for other exposed residential and worker populations. The people who could be affected by indoor dust levels were cleanup workers and residents returning for items and assessing damage. Exposure period 4 was the exposure primarily due to the emissions directly from the smoldering fires, which burned until December 20, 2001.

The patterns of exposure described here for the complex mixture of dust and smoke and the types of particle exposure that would have occurred in the hours and days subsequent to the attack would be consistent with the incidence of health outcomes in the firefighters as reported by Prezant et al. (2002). These would have been augmented by the levels of gases released during the fires. Included would be the incidence of WTC cough and bronchial responsiveness in the firefighters, and the time during which they started working at the WTC site. This was amplified by Prezant et al.'s observation that few, if any, of the firefighters wore respirators or wore them properly during the rescue operation. Thus, first-responder firefighter exposures to the initial dust and smoke would have been very high, since these individuals were on shifts in excess of twelve hours during rescue operations. The patterns of higher to lower exposures as presented in the Prezant et al. article could have easily been placed adjacent to exposure periods 1, 2, and 3 described for WTC generated or resuspended dust and smoke. The time intervals spent by the firefighters postcollapse at Ground Zero were described in the self-reported questionnaires, and results indicate that first groups of firefighters would have had the highest and longest direct exposures to coarse particles and fibers, and particles with a high pH. The medium-exposure firefighters group would have been primarily affected by resuspension of the dust and smoke initially produced on 9/11, and smoke generated during the rescue activities on September 12 and 13, which also preceded the first rain and vacuum cleanup of Wall Street. These exposures would have been lower since the levels of particles and gases would have declined. The low-exposure firefighters would have had reduced exposures to resuspendable coarse particles and high-pH particles because of the reduction in available dust caused by the rain. Exposures, however, would have continued for a relatively long period of time. In all three cases, the composition of the exposures would have been documented by the bulk of the initial mass, high-pH coarse particles that contained materials like glass fibers, cement particles, and soot, and direct inhalation of smoke from the smoldering fires (continuing). However, the levels would have decreased quickly and approached background levels after the first week or two. For the gases released during the fires, the decreases in concentration would have closely followed the intensity of the fires. Thus, the magnitude of the concentrations, though not measured would have decreased more rapidly than the particles (no resuspension). Thus, the levels would have affected those firefighters most significantly during exposure period 1 and then less significantly during exposure period 2. Afterward, the levels would have decreased rapidly and concurrent with the extinguishing of the individual fires.

In the discussion and conclusions, Prezant et al. present a case for upper respiratory symptoms and distress that were manifested by the WTC cough, bronchial responsiveness and nasal congestion, and gastro esophageal reflux. The original exposure data were inconclusive about the pertinent exposures. The actual materials of concern were not initially determined in specific samples, and initial data focused either on asbestos or some single fraction of the mass. Thus, coupling the Prezant et al.'s results to the Lioy et al. results provides a strong argument for the types of acute exposures that could have led to the acute responses seen in the firefighters. The evaluations indicate nontypical pollutants that were generated and were a cause for concern for acute health effects. Furthermore, the size range was not typical of concern in the general environment, those greater than 10 µm in diameter [supercoarse particles], which would lead to upper respiratory deposition. Also, the agglomeration of small particles and longer fibers would lead to secondary sites for release as time went on. The very high mass loadings and gases released during exposure periods 1 and 2 were a unique complex mixture of building materials and smoke that was formed by pulverization, or burning jet fuel and materials before or after the collapse. More than 99 percent of the particles were inspirable particles in the particle size range above 10 µm in diameter. The gas concentrations were unknown.

In the end, the magnitude of the exposures that have been or will be related to WTC health effects will never truly be known, although the government, through the ATSDR WTC registry, has identified quite a few of the individuals who were there during those first two days. I never registered but should have. For those who have documented presence at Ground Zero during the first two days postattack and did not wear respirators, the analysis of the registry data, even with crude exposure indicators, will probably give us an opportunity to understand the effects that resulted from the WTC aerosol exposure. As a consequence, the number of victims of the attack has expanded from approximately three thousand dead to include over six thousand injured responders. This point has never been clearly put into perspective and should at some point in time. Basically, with the rush to rescue without adequate personal protection and the lack of knowledge of the potential effects of the WTC dust, the actual number of victims of the attack tripled during the period while rescue efforts were proceeding at Ground Zero.

The registry has issued a few reports, and the conclusions appear to be consistent with the work of David Prezant and others at the FDNY that indicated that the highest frequency of effects occurred among people, rescue workers, and the general public at Ground Zero during exposure periods 1 and 2. In 2008, around the seventh anniversary of the attack, the ATSDR registry investigators published a summary of the results. Using the registry data, it reported that of the 71,437 who met the minimum eligibility requirement, between "3800 and 12,600 adults experienced newly diagnosed asthma, and 34,600–70,200 adults experienced post traumatic stress disorder (PTSD)" (Farfel et al. 2008). Most of these individuals were either present in southern Manhattan near or inside the WTC on 9/11, and many others were rescue or recovery workers. Howard Kipen described his clinical work on the health issues in the years that followed the attack as follows: "I did not then anticipate that as occupational and environmental clinicians we would be so busy almost a decade later caring for and researching the physiological and psychological consequences in response workers who did those jobs" (Howard Kipen, pers. comm.).

In 2009, the Mount Sinai School of Medicine and the New York City Fire Department published follow-up studies on their patients that were associated with the aftermath. In each case the data is clear that a certain percentage of the individuals continued to have some sort of respiratory symptoms. Furthermore, those who came during exposure periods 1 and 2 had the most, especially among those who did not wear respiratory protection. An interesting comment in the FDNY conclusions was "In most large disasters, exposures may be unavoidable during the rescue phase, but our (their) data strongly suggest the need to minimize additional exposures during recovery and cleanup (reentry and restoration) phases" (Webber et al. 2009). Will we ever be able to pinpoint the exact compounds that caused respiratory difficulties and other physical effects? No.

Will we be able to document the synergism between the gases and particles inhaled and the inhalation health effects? No.

However, I do think that we can do a better job in reducing exposures during rescue efforts in the future. That requires much better training about the consequences of exposure and the need to protect oneself at the same time you are attempting to rescue others.

People were exposed to the WTC aerosol, and that aerosol was a complex mixture of gases and particles, a mixture that I hope we will never see again or have to measure again. The fact that we detected many materials and can try to approximate the gaseous component will eventually allow some Ph.D. student to paint a picture of the nature of the severity of the impact of such a mixture on the human respiratory and digestive system. However, because of the nature of toxicological testing and because of the fact that we do not have an exact idea or, for that matter, much information on the composition of the gaseous material, we will never have a full understanding of the toxicity of the WTC aerosol. We have a clear understanding of the WTC dust, and some information on its toxicity, but is that sufficient? No.

SOME FURTHER OBSERVATIONS ON THE DUST

Eventually David Prezant and I collaborated on one WTC-related study. It was designed by Dr. Elizabeth Fireman, University of Tel Aviv, Israel. She wanted to use a technique called inducing sputum (inhalation of hypertonic saline solution to obtain a mixture of saliva and mucus that is coughed up from the respiratory tract and examined, in this case, to aid medical diagnosis). It was used by Dr. Fireman's team to compare the material that was present in the lungs of Israeli firefighters to the material present in the lungs of WTC Firefighters from NYC. In addition, they used some of our samples and results to compare with the electron micrographs taken of the sputum samples. One interesting point was that the sputum measurements on the New York City firefighters were made ten months after the WTC collapse, thus the results would give some information on the firefighters' ability to remove the WTC dust from their lungs by natural lung clearance processes. Our results showed that the most highly exposed FDNY personnel had much greater inflammation as indicted by tests for neutrophils (a white blood cell with sacs of enzymes that help kill and digest microorganisms). The FDNY personnel measurements indicated larger and more irregular shaped particles similar to the materials we had found in the three WTC dust samples. Thus, we had information that some of the material in the WTC dust actually penetrated deep into the lung and stayed there for ten months. How long it will take to clear is unknown; we did not get the necessary funds for a good follow-up. Another opportunity lost.

All of the chemical and physical analyses completed on the three outdoor samples were validated by the end of October. At that time, however, we will still waiting for permission to obtain indoor samples from our selected locations downtown. The efforts of Lung Chi got us permission for sampling a number of locations, but we still did not have the approval needed to go into the zone of exclusion. The effort to obtain approval continued to be fruitless throughout October.

About that time, I felt we should start writing a report on the WTC dust samples. After some discussion with Lung Chi Chen, it was clear that these reports should be scholarly scientific articles. We identified about three or four possible articles with shared authorship. However, I had some real problems beginning to translate my task, writing about the composition of the WTC dust samples, onto paper. It was not that I did not know how to do it, but it was how to write it and keep the content to just to the facts. We already had some difficulty with submitting abstracts to meetings because of sensitivities for the victims, and I again wanted to be very careful about how this science would be received by readers and members of the general public. In any case, the stumbling block to writing was me. Gathering together of all the data and selecting the journal style took about a week. I was ready to start, but then nothing happened. I knew what to do, but still nothing happened. I even called the editor of *Environmental Health Perspectives*, and asked if the journal would be interested in it for review since the article was not a typical submission. The editor said yes, but that assent still did not get me in the frame of mind to write. Over the next week still nothing happened. Fortunately, I had the World Series for diversion, the Yankees vs. Arizona. The throwing of the first pitch by President Bush was electric, and the late inning comebacks by the Yankees were great. Arizona finally prevailed, but it was a bit of excitement and entertainment during the time of great uncertainty and stress. The next week I had a meeting in South Carolina on exposure science, the society's annual meeting, so I stopped thinking about writing for a few more days.

At the International Society of Exposure Science (then Analysis) meeting, many attendees started to talk about the data from the aftermath. However, most of the local attention was focused on bioterrorism and the anthrax terrorist attacks. I was interviewed by a local TV channel, and to my surprise they focused on the anthrax attacks, not WTC. Did we get it wrong? Had people already begun to forget? Did they not understand that we were at war? I guessed that when you left NYC, the world did look different, or there could have been the desire to think of other things. Thinking about the times a bit more, could I have been wrong about the guy jogging toward Ground Zero on the 17th? Was he just trying to hold on to a narrow corner of his world to survive the trauma of the days immediately after 9/11?

At the annual meeting, we talked a lot about terrorism, and there was a well-attended impromptu briefing set up at the meeting by members of EPA and NIEHS staff to discuss the current status of the environmental and occupational responses to the aftermath. I think this was a very informative and thoughtful session. Upon returning home we were still waiting for approval to go downtown, but I finally got an inspiration from a likely source—Jeanie. Over dinner one evening I was lamenting that I just could not write. She looked at me and simply said, go watch the DVD on Pearl Harbor—*Tora Tora Tora*. I had seen the more recent version, *Pearl Harbor*, in the movies with my son Jason the summer before, but that was not yet on DVD. That return to the past helped. This series of steps, and more, may seem a bit odd, but I think that just like in sports you need to get yourself ready to do your best or "game on."

THE QUEST FOR INDOOR DUST

As I was reviewing the results and preparing a manuscript, the indoor issues in buildings around Ground Zero began to mount. There was no true plan in place for cleanup. Some guidance had been given by the city of New York, but there was no coherent plan. Our team continued to be frustrated because we were unable to get into any buildings, and as the days passed, my usual optimism was being challenged. Thus, as the calendar turned to mid-November, sampling indoors still had no options. As people were going back to work, they encountered days of odors in the office from the now smoldering fires. Thus, the indoor issues continued in southern Manhattan.

To my dismay, January 2002 was the first time we found out that the ATSDR had been working with the city and had conducted sampling in thirty homes in Lower Manhattan and four background locations above 59th Street from November 4 through December 12, 2001. What a missed opportunity! Again, here we were, NIEHS centers funded to complete thorough analyses of WTC samples, and there was no request to join. It was not as if we were unknown entities. By that time we were well represented in the community from the ongoing activities of NYU and Mount Sinai.

This aside, the biggest loss was in the kinds of materials presented in the reported results. The researchers' entire focus was on the building materials. The analyses were good because they did measure vitreous fibers, synthetic fibers, and gypsum (irritants), which was a refreshing addition to asbestos and silica, which were also measured. However, they did not report any levels of lead, which we had provided as a chemical of concern, and reported few measurements of pH, a variable of health interest, in early October. They also took air samples but did not report the mass concentrations. These missed opportunities could have been avoided by having an inclusionary approach. Clearly that lack of inclusion was a problem, since the final report included reference to the excellent work completed by the USGS, hence the addition of vitreous fiber. However, the ATSDR totally ignored the report of our team published in July 2002, which had been available online for about three months and had been reported on by one or more team members to the community and agencies since October

2002. I am still annoyed by the lack of inclusion of the NIEHS center investigators.

In contrast to the ATSDR, and which may be a surprise to some, but not to me, the EPA scientists engaged with the NIEHS supported investigators on the dust issues, and we were very helpful to each other. However, I do not feel that these dedicated scientists from the Region II, New York City office, Research Triangle Park (the National Laboratory), ever got credit for much of their work. As a simple example of their efforts, it was through Dr. Dan Vallero of the Office of Research and Development in the EPA that we finally received access to Ground Zero for the collection of indoor samples. The clearance obtained by Dan came from the most unlikely of sources, the U.S. Coast Guard, which was overseeing many operations. The buildings selected for sampling were obtained through the work of Lung Chi Chen, and the managers of the buildings were still on board for our field team to collect samples.

The day our team of scientists from NYU and EOHSI finally went to New York to collect samples was November 19. It was again a relatively warm day, which is unusual for late November. This time Lung Chi and I wanted to leave nothing to chance. So we decided to go along with our field team to collect the samples. Thank goodness we made that decision. Our indoor sampling team met at NYU's downtown hospital and then traveled to Ground Zero (figure 9.1). This was the first time since September that I had been physically at the site, and the conditions had changed in the local area. Ground Zero was now an organized work zone continually frequented by diesel trucks and other vehicles for debris removal. These would be the primary source of local air pollution for the next six months. The entire area also was cordoned off with fences and security personnel.

Although we had been given assurances that the security personnel at the site had been told that we would be arriving, we were met, as our field team in October, with resistance from security to allowing us to enter the area. This time we did have the necessary documentation, and we were able to convince the guard to take us by cart to the coordinating center. There we received our badges, and we were finally able to visit the places that agreed to let us take indoor samples.

The first place we visited was located on Liberty Street. We were fortunate to have full access to the building, and we took samples from

Figure 9.1. Indoor sampling team: (from left to right) Paul Lioy and Vito Iacqua from EOHSI, and Lung Chi Chen, John Gorczynski, and Michaela Kendall from NYU. Photo taken by Dr. Lung Chi Chen, NYU Institute of Medicine. Reprinted with permission.

condominiums located a number of stories above the ground. The samples collected at our second location on Trinity Place were from accessible rooms in the building and had observable levels of WTC dust on surfaces. We did not have any problem seeing WTC dust-laden surfaces, and some examples are provided in figures 9.2a and 9.2b. The most striking image of the entire trip, and one that we published in the *Environmental Health Perspectives* article in 2002, was the teddy bear in a high chair with its table covered completely in WTC dust. It was a vivid picture about the end of innocence.

Each location, however, did present challenges to sampling. The biggest obstacles were as follows: each unit we entered was untouched since the day of the attack, and the conditions were the same or worse (e.g., decayed food, rodents, broken windows, etc.), each unit had different levels (depths) of dust present on the sampled surfaces, and each unit presented safety hazards.

Figure 9.2a. An uncleaned indoor location that our team visited on November 19, 2001. Photo taken by Dr. Lung Chi Chen, NYU Institute of Medicine. Reprinted with permission.

There was another "New York moment" that day. During our sampling visit to the building on Liberty Street, we were given access to all units except one. That unit was an oasis in a sea of WTC dust. Apparently, one of the condo owners had already contracted to have their unit cleaned professionally. The owners were going through the process of wet and dry wiping and vacuuming, and the unit looked very clean. However, I wondered while looking through the plastic that covered the doors, how long until the owners are able to return? Since the rest of the building had not been cleaned, there was WTC dust everywhere, and many units were severely damaged. In addition, what safeguards would be required to ensure that the unit will not be refilled with dust during the cleaning and the repairs of the other units and the hall and stairway? With no clear policy to deal with this type of question, it would be asked frequently in southern Manhattan. The good news was that with professional cleaning and restoration the residence looked like it had never

Figure 9.2b.   A teddy bear covered with WTC dust, November 19, 2001. Lioy et al. *EHP* 110 (2002): 703–714.

been touched by the WTC dust released during the collapse, although it had been severely impacted during the collapse.

The visits to these indoor locations solidified my concerns about indoor issues, but it also reminded us that the loading of dust on surfaces was just as important as the concentrations per unit gram of dust. My reasoning was based on the way in which we approached the lead issue, surface loading as the potential exposure metric, and the confusion that reigned about the meaning of 1 percent asbestos in the WTC dust and indoor cleanup. It was not just the concentration of a material in each gram of dust; it was the total amount present on the surface, the grams of dust on the surface. As indicated in figures 9.2a and 9.2b, a lot of material could be spread on surfaces throughout a home.

We returned to the laboratory that evening, and the indoor samples were stored and subsequently separated for analysis. All samples were shared between Lung Chi and our laboratory. The samples were weighed, then packaged by my staff, and sent out to the same laboratories that completed the analyses on the outdoor samples. In addition to the samples we collected on November 19, over time I also received a few samples from local residents, which were also prepared and submitted for analysis.

To be consistent with our previous work, the samples were analyzed for the same extensive group of inorganic materials and elements, organic compounds, and pH. By this time, it was clear that one of the first measurements that needed to be made, and one that could reasonably identify the WTC dust, was pH. The high alkalinity of the dust was one of the drivers in the post-WTC health effects being observed in the workers, and it appeared to be influencing longer-term inflammation of the upper respiratory system that was first noticed as the WTC cough.

For all of the samples collected by our team from within the two buildings near Ground Zero, the pH of each sample was above 7. This fact attested to the alkaline nature of the dust and provided continued motivation to cleanup the WTC dust professionally. In contrast, the residential samples provided by residents from North Moore Street and South End Street were below 7. A Liberty Street window sill sample obtained from a resident six months after the collapse did show the presence of WTC dust, which indicated that WTC dust was still present in isolated outdoor locations at that time. In addition to pH, two elements that appeared to be related to the WTC were the concentrations of titanium and lead, but since they are very common chemicals, it would be hard to call each a tracer or signature for WTC dust.

The morphology of the indoor dust was similar to that of the three outdoor samples with the results from all eighteen samples showing the presence of glass fibers, calcite, vermiculite plaster, and charred wood fragments. Chrysotile was present at levels of less than 1 percent, which was consistent with most of the outdoor values. Additional materials, such as starch and cotton fibers, were found in some indoor samples. These probably were already present on the floor or other surfaces in the residences or spread by rodents during the sixty-seven days that the units were unoccupied and open to the downtown environment. In appendix A, I have provided some details on the results from the indoor samples taken at the Liberty and Trinity Street buildings. More can be found in the Yiin and Offenberg papers cited in the bibliography.

The organic compounds present in the indoor dust samples had a slightly different character than the compounds observed in the original outdoor samples. Since the indoor samples had been collected approximately two months after the attack, the more volatile organic compounds deposited on the dust particles from the jet fuel and other fuels stored at the WTC had evaporated from the particle surface. Some of the alkanes were still present as well as compounds with lower volatility (slower to evaporate). Included were some phthalates and PAHs. Dioxins were present at levels similar to the outdoor dust. So, although the general character of the dust remained the same, materials that could be transformed or evaporated were gone. As with some other environmental issues, the character of the problem changed over time, and the WTC dust would continue to change character the farther the analyses completed were from the original date of collection. This is an important point to remember for any type of environmental and many other forensic types of sampling. A point for the future would be to analyze dust samples soon after collection because some of the compounds, including important tracer materials, can degrade (change or disappear) over time. Furthermore, this problem would cloud the results of any analyses completed on the WTC dust today. That is why I will only give dust out now as historical artifacts or as educational research material.

The PAHs were again analyzed in detail by John Offenberg, and as with the outdoor samples, the sum of thirty-seven PAHs measured were about 0.04 percent of the mass. However, John took a next step of seeing whether he could identify a WTC PAH profile that could provide a fingerprint for WTC dust. As I said earlier, a few of us could easily identify the dust when we saw it. However, just like in criminal forensics, to establish the potential cause-and-effect relationship, a fingerprint is needed. John made the first attempt by working toward identifying a WTC PAH profile for the indoor and outdoor WTC dust samples. It did show some promise, since the levels of specific PAHs became more irregularly distributed in the potential fingerprint profile after a unit was cleaned up. However, in further tests, it could not be used as a fingerprint that could define whether a unit or residence was clean enough for rehabilitation. Other variables would be needed to provide a stable fingerprint. The reason why PAHs failed to be a distinct fingerprint or signature is that there are many sources that during the combustion process produce varying quantities of PAHs in the emissions. The other sources will distort any fingerprint and, in terms of science, will increase the variability of the components of the fingerprint. In simple terms, it will be difficult to pick WTC dust out from other material that can be burned in a residence, such as frying of meat or tobacco smoke. However, in the winter of 2001 to 2002, we were not in a discussion of fingerprints or signatures. We were still trying to identify the indoor problems, and it was becoming clear that surface loadings indoors would be a significant cleanup issue.

One final feature of the dust found indoors was substantially different than the outdoor samples: the sizes of the particles that were present in the samples. For the indoor samples, the particle size range was more compact. There were fewer particles in the less than 2.5 µm size range (fine particles), and there were fewer particles greater that 53 µm in diameter for all eighteen indoor samples except for one. That particular sample had a larger amount of clumped material than the other indoor samples. It is hard to know exactly the reason for the tightening of the size distribution of particle sizes, but it probably has something to do with the larger particles being filtered out of the air as the dust penetrated around windows and doors, when the pressure caused by the collapse of the towers accelerated the dust into buildings. WTC dust going through broken windows would be unaffected. Also, as the WTC dust moved from the rooms closest to the towers to more interior locations, the size changed because the larger, supercoarse particles would deposit first.

Ultimately, these scientific curiosities meant nothing in terms of cleanup strategies. From our results it was clear the cleanup issues were immense, and the details of the approach needed were still not thoroughly understood. For example, questions that arose included these: What furnishing and personal items to keep and what to discard? What constituted a clean residence? The first question was an immediate issue, and the second was the long-term problem that I had identified to Mike Gallo during our E-Team visit on September 17. Mike agreed with my assessment that day. However, in terms of health outcomes that could be reported by indoor cleaning crews who did not wear respirators, there would be some differences in exposure when compared to the workers at Ground Zero during exposure period 1—namely, (1) few, if any, high concentrations of volatile gases associated with the emissions fires were being transported indoors after the first week; (2) there were less supercoarse particles to be resuspended in the air, but (3) the frequency and length of exposure could be longer because multiple buildings could have been cleaned by multiple individuals.

10

TALKING AND WRITING:
TO WHAT END?

After the indoor sampling was completed, Lung Chi and I were interviewed by Discovery Channel (Canada) in early December. The reporter was very thorough and focused on the dust. The interview aired in January in Canada and many other places around the world, but not in the United States. This was another missed opportunity to explain the WTC dust issues to the NYC community. By mid-December, the fires were reduced in number and size, and they officially came to an end on December 20, 2001. The data that had been collected from the ambient air monitoring sites around Ground Zero were showing levels that were typical of the "normal" NYC air. For example, the intensive EPA monitoring network in southern Manhattan found fine-particle concentrations that ranged between 5 and 45 µg/m³ at all its sites from December 20, 2001, through the end of the year. A much smaller network of measurements established for particles less than 10 µm in diameter (inhalable) did not report values above 30 µg/m³ (i.e., 30 µg of material in every cubic meter of air in the local atmosphere) during that time period, and the matching fine-particle measurements never went above 20 µg/m³. Thus, when these levels were compared to the twenty-four-hour ambient air quality standard for the fine and inhalable particles that existed at that time, there were no excursions above either. The twenty-four-hour health standards for fine particles and thoracic inhalable particles were 65 (today it is 35) and 150 µg/m^3, respectively.

For the gases, the most familiar was the carcinogen benzene, and by December 20, there were no reported values above 1 ppb, which was essentially urban background. This was in contrast to early October when the benzene values could exceed 10 ppb. This level has been seen in homes with attached garages and other sources of benzene. In December 2001, the levels of most volatile organic compounds were at or near 1 to 2 ppb. On one day, December 31, the ethylene benzene level reached 43 ppb, but this is still not a WTC concern, since this compound is emitted by fuel oil or gasoline. The temperature on this date in NYC was below normal; therefore, higher fuel oil use would be anticipated for residential and commercial heating.

Just prior to Christmas, I was interviewed by the *Wall Street Journal* (*WSJ*). Christmas was quiet that year for obvious reasons, but I was busy writing and sending out drafts of the article to my collaborators for review. On December 26, 2001, the *WSJ* came out with a page one article by Mark Maremont and Jared Sandberg that I felt set part of the tone that would keep me involved with the WTC dust for the next few years. Very factual, it basically put the science and the uncertainties into a reasonable perspective. The article stated, "Tests say air is safe, but some people feel ill near Ground Zero," and "much remains uncertain" (Maremont and Sandberg 2001). The *WSJ* clearly indicated that there was general agreement on the workers being the highest at risk and that the predominant problems appeared to be respiratory irritations. Many health scientists were beginning to understand and talk about cement and glass fibers. For nonrescue workers and emergency personnel who were at Ground Zero during exposure periods 1 and 2, a quote from George Thurston of NYU sheds some reasonableness to the situation as it was perceived at that time. As quoted in the *WSJ* article, "The government is right that otherwise healthy people are not going to end up dying or in the hospital . . . but some of the dust and chemicals kicked up by the collapse . . . turned out to be more irritating than we thought" (Maremont and Sandberg 2001).

Our paper on the WTC dust was submitted to *Environmental Health Perspectives* on January 15, 2002, just four months after the attack. I was pleased with the efforts made by all authors to make contributions and edits, especially over the holiday season. However, the submission of an article was not sufficient. Considering the issue at hand and the time it takes to review an article and then, if the article is accepted, the time until it appeared in print, we needed to get the information out quickly. However, I was still very reluctant to submit an abstract to a scientific meeting at the time, and did not. So, at EOHSI, the acting director, Dr. Robert Snyder (the same person who landed during the attack on the Pentagon), Mike Gallo and I decided that it was time to give a public presentation of the data at our institute.

We chose February 15, 2002. This was about a week after the *St. Louis Post-Dispatch* and *USA Today* published front-page articles about the environmental issues. The title of the *USA Today* article was "Anxieties over Toxins Rise at Ground Zero." The timing was right, and I was interviewed about two days before the presentation by Ted Sherman of the *Star-Ledger*, a New Jersey paper, published in Newark. On February 16, the newspaper published a page two article, with pictures, that provided an excellent summary for the general public about my presentation.

Numerous reporters, scientists, members of the general public, and government officials came to the event hosted by EOHSI. The estimated total in attendance was over three hundred people for an afternoon seminar presentation at the Robert Wood Johnson Medical School East Lecture Hall in New Jersey. There were lots of questions, but this time I had some answers on the types of exposures that probably occurred as a result of the inhalation of dust and smoke. Many of those present were very interested in the high pH (alkalinity) of the WTC dust and the glass fibers. Lung Chi and I had previously mentioned this point at our October meeting, but at that time no one reported the content of our presentations.

During the days before and after my presentation, the indoor issue became more complicated. The 2003 GAO report stated:

> Initially, building owners were held responsible for cleaning up their own building, including interior and exteriors. According to New York City officials, the issue of funding the cleanup of privately owned buildings was discussed with FEMA and EPA; and the initial federal position was that the Stafford Act (the implementing statute for the federal response plan) did not provide direct funding to New York City for this clean up.

New York City officials said . . . that they informed the federal agencies that the building owners would be responsible for the cleanup of their buildings and the federal agencies agreed with this position. Under this arrangement, owners of rental units were responsible for cleaning apartment walls, ceilings and floors, common areas . . . and heating, ventilation and air conditioning (HVAC) when deemed necessary as explained in guidance provided by New York City. Renters were responsible for cleaning personal belongings. In resident-owned condominiums, residents were responsible for their units, while building owners were responsible for cleaning common areas and HVAC. (U.S. EPA Region II 2003)

There are many aspects and opinions about the above comments and these also can be found in the same GAO report. To most scientists, this description would sound like a reasonable and well-balanced plan. Unfortunately, the indoor issues were not easy to handle. More importantly, the strategies were not easy to implement. In February 2002, the EPA assumed the chair of a multiagency task force on indoor issues. The final result was the development of a "Test and Clean" program for the residents of Lower Manhattan.

Using the results from the dust analyses completed by our team, and data collected by EPA from the landfill in Staten Island, the ATSDR, and other sources, the EPA completed the thankless task of establishing benchmarks for indoor cleanup without the benefit of many years of research and evaluation that it usually takes to develop national standards or benchmarks/guidelines. Who would have thought to test WTC dust? This term was nonexistent before 9/11, an important point to remember. Time was of the essence, and the EPA and others had little time available or data to use to establish guidelines.

The chemicals of concern that were reviewed by EPA included a number of the materials that we detected in the dust. The ones selected for use by EPA were lead, PAHs, and dioxin in settled dust, as well as indoor air. In addition, EPA researchers chose asbestos, alpha quartz, and fibrous glass (vitreous fibers). They used these compounds since each would be present in resuspendable dust and act as representative components of the different types of materials that needed to be collected by a variety of methods and devices for indoor cleanup, a wise decision.

My reasoning was based on the fact that the agency selected a number of materials that can cause either long-term or short-term health outcomes. The materials selected included both fibrous and nonfibrous particles, and all were present in the resuspendable dust. The materials ranged from being present in trace quantities (lead) to a large fraction of the mass (fibrous glass) which would help to truly determine the effectiveness of a cleanup. Furthermore, without definite cleanup numbers, such as PALs, the EPA approach would provide a sound engineering approach to achieving "how clean is clean" given the complexity of the composition of the WTC dust.

Some of the materials used for evaluation did not have a health benchmark. These included components of cement and building materials that could be present in the air and dust. For example, asbestos in dust, vitreous fibers (later to be called man-made vitreous fibers [MMVF]), calcite and gypsum in air and settled dust, and total settled dust. As in any well-designed field study on effectiveness of exposure reduction or risk reduction, samples of dust were taken before and after, and analyzed for the original chemical of concern.

The effort by EPA to establish the methods to be used in the cleanup of residences was called the Residential Confirmation Cleaning Study, and a two-volume set described all the results. I will summarize the most pertinent results for the WTC dust removal.

To test the eleven cleaning methods, the EPA selected two methods for each of thirteen residential locations and five methods for each of two commercial locations. The comprehensive and quickly completed investigation provided the EPA with the ability to define the range of cleaning repetitions needed for achieving clearance in an indoor cleanup program. The work provided much better information to those individuals who would want to have a professional or a nonprofessional cleanup of their residence. However, it should be remembered that most plush materials and furnishings would have been very difficult to clean.

In the end, the EPA found that at least two and, at times, more cleanings were necessary to achieve the chemical benchmarks for attaining a clean residence or business space. However, the most important point was that EPA was able to truly provide technical guidance on how to clean buildings of WTC dust. The bad news was that we were so ill prepared as a country for 9/11 that it required months of assessments and evaluation to reach these recommendations, and provide them to residents and building owners. Unfortunately, good science and engineering were lost because of the lack of trust by the community, and the constant second-guessing of the science by many members of the general public, elected officials, and the press. Someone should at least memorialize this work to prevent the need to "reinvent the wheel" sometime in the future when another catastrophe involving some kind of dust or settled material arises.

Our first paper on the WTC dust was accepted by *Environmental Health Perspectives* and was published electronically in April 2002 and in print in July 2002. This was remarkably fast for most scientific papers, especially since this was an article that was relatively complex in content with multiple authors. The commitment of all involved to getting the job done, including the journal and the reviewers, made the difference during these difficult circumstances.

As time went on, the WTC dust was being removed from buildings because of cleanups, professional or nonprofessional. In June 2002, I attended a meeting at the New York Academy of Medicine that was held in collaboration with the EPA regional administrator, Jane Kenny, to discuss the EPA plan for a cleanup program. We also received an update on the EPA monitoring efforts and listing of the chemicals of concern in the WTC dust. It was an interesting session, and some good science was discussed. The question was, Did science matter?

The EPA voluntary Test and Clean program started during the summer with the first cleanup taking place on September 12, 2002. One would think that with all the efforts in science and engineering that went into designing the protocols that the issue would be heading toward a reasonable conclusion.

Many of you would not have followed the development of the program; so let me fill in some blanks. Besides the EPA Residential Confirmation Cleaning Study, the EPA completed a Background Dust Study to help define the typical background dust levels of the Chemicals of Concern in the New York metropolitan area for the cleanup program. The EPA published a report called the "Selection of Chemicals of Concern" that helped to define the targets for consideration for the indoor cleanup. These were all major tasks that had never been considered by the EPA or any other agency prior to 9/11 for residential dust. Again, one would think that the community would have been thoroughly engaged and eager for the cleanup of the indoor residences.

As we entered the fall of 2002, my confidence was up; so my son Jason and I took our overdue business/personal trip to the home of our ancestors—Italy. We had a great time in Rome and in Capri, which is just north of Bologna. Later that fall, my optimism about WTC dust restoration issues was crushed again. The registration for the EPA Test and Clean program was completed by the end of 2002; however, the distrust and fear that had emerged about how the indoor WTC dust issue was addressed remained, and criticism of the program was visible within the community.

2003: A TURN OF THE PAGE

Entering 2003 was both positive and negative for the WTC dust issues. Yes, the voluntary Test and Clean program had started, but too many people were asking the impossible: "Please put Humpty Dumpty back together again." This was a problem because the towers were gone, and everyone's lives were changed forever, no matter what the wishes of the community were, and no matter how many agencies could get involved. Science and policy could only move on and attempt to do things better and prevent another attack. However, as you will see later, some of these issues remain in play today.

By the beginning of 2003, our group and others had published a number of articles on the WTC dust and the air quality in NYC. One came out just before Christmas 2002. I was quoted in the newspaper again on the dust. This time it was based on an article written by Offenberg and a number of colleagues on the PAHs. There was not much new to say, but the WTC dust was still news.

The publication of David Prezant's work on the firefighters had a major impact on the way worker's health issues were being addressed. The ATSDR had established a World Trade Center registry in 2003, mentioned previously, for those who thought or knew that they had been exposed to WTC dust and aerosol during those early days after the event. As 2003 began, the indoor issues and the respiratory effects in those who had not worn respirators were not going away. In fact, the GAO was well on its way toward its 2003 report on the response, misnamed "EPA's response to . . ." when it should have been "Government's response to . . ."

In a complex response, where no one was given full authority to lead any aspect of the response, it is better to examine the entire process. Then one needs to determine what has to be done to make things run smoothly the next time. To select one agency leaves others off the hook and does not breed better communication and collaboration. Communication on many issues related to homeland security has improved significantly over the past eight years; however, a bipartisan report in 2008 indicated that there is still a lack of integration of many U.S. programs.

The WTC dust was still front and center during the winter of 2002 to 2003 when EPA issued a draft report on a risk assessment for the World Trade Center. The report focused primarily on long-term health effects. It did not include acute effects or how inflammation could eventually lead to longer-term health consequences. It focused on the more pre-9/11 issues of long-term health effects, and asbestos was a major topic. WTC dust, however, grew as a topic of discussion as more and more workers who were at Ground Zero during those first few days were being diagnosed with upper respiratory and stomach-related problems. Again, more confusion was displayed that would not fade away. A paper based on the report was eventually published in 2007 by M. Lorber. He did a good job summarizing issues related to both acute and long-term health effects. However, we never did see a final version of the draft EPA report.

A SIGNATURE AND THE LACK OF TRUST

With the characterization of the WTC dust done, you might think I was at the end of the story, but I was not. We entered a new phase of the WTC dust story, but now it related to specific activities that attempted to continue restoration and rehabitation of downtown Manhattan, and efforts to continue to reconstruct the exposures at Ground Zero. During 2003, the EPA was completing its Test and Clean program in downtown Manhattan. By the time the sign-up deadline of December 28, 2002, arrived, the cumulative totals were as follows: 730 homes to be tested only, and 3,436 to be cleaned and tested. FEMA provided $60 to $80 million for the voluntary program.

It was the next chapter in the WTC dust story when communication with the local community truly broke down. Furthermore, the continued presence or absence of WTC dust became part of the discussion during the attempts at restoration that would lead to rehabitation of the areas in downtown Manhattan.

The indicator of the "presence of WTC dust" used as part of the EPA Test and Clean program in the "testing only" phase was resuspendable asbestos fibers. The tools used to test for asbestos, in EPA's own words, were the

> *modified-aggressive testing* approach which is designed to simulate the room-air movement you would expect in normal living conditions. Twenty-inch fans (one per 10,000 cubic feet of room space) are run for the duration of the *indoor air* sampling. . . . *Aggressive testing* was also offered and it was conducted in the same way; except that, prior to the beginning of air sampling, a one-horsepower leaf blower is used to direct a jet of air towards the corners, walls, fabric surfaces and ceiling to dislodge and resuspend fibers and dust. Aggressive testing is usually used following an asbestos removal project in an unoccupied building. The aggressive method creates room-air movement that is much greater than you would expect to occur in a home. In an occupied residential environment, particularly one with carpeting and other fabrics, it is possible that the vigorous air flow will loosen so many fibers that the air samplers become clogged. If the air sampler gets clogged, the test becomes invalid. (EPA, Cleaning & Testing Fact Sheet, www.epa.gov/WTC/factsheets/cleantest.html)

From the standpoint of dust in general, the second approach intrigued me. If I were the home owner, what would I want done to my residence? Aggressive or modified aggressive testing of the residence? As we all know, dust has been around with us forever, we have devices of all types and varieties available to pick it up from floors, tables, and other surfaces. In addition, some people have very clean homes, and others do not. The chemicals in the world around us do get distributed into various locations, and some tend to accumulate around the home (especially under and on top of refrigerators). In fact, the best way to determine the influence of personal and environmental activities in a residence on exposure to toxicants that have relatively long lifetime is to examine the dust found in undisturbed areas of a home or in the carpet. Thus, under normal or reasonable circumstances, dust does not move much unless you move it by a vacuum cleaner or a broom. By disturbing the dust, as was done by EPA in the Test and Clean program, there was a strong indication of resuspendability of WTC dust present on all the surfaces in the home, and a demonstration of the maximum potential for inhalation of WTC dust by the residents. So to me this was a reasonable approach for the "testing only" part of the program as long as the dust was WTC dust. Remember, asbestos can be present in NYC dust without the WTC dust. Therefore, without a signature or fingerprint for the WTC dust, one must make assumptions about where the asbestos came from. In many cases, if there had been WTC dust present in large quantities, the testing of a residence would not give false negatives because you could probably see large quantities of visually identifiable WTC dust in the residence. The approach would be acceptable as long as the residence was eventually cleaned up.

During the 1990s, I was involved with a study that evaluated the collection efficiency of vacuum cleaners used in homes. This was part of my excursion into how kids get exposed to lead and chromium dust, and how contacts with dust can result in increased levels of these elements in the blood or urine. The latter would be observed by reviewing the results of biological marker measurements made for the lead or chromium concentrations in blood and urine samples, respectively.

The vacuum cleaner tests revealed that many of the vacuum cleaners available at the time primarily picked up very large materials, the bulk of the mass, and spread the smaller particles around a residence. The vacuum cleaners were not designed for the purpose of picking up smaller particles. You have probably heard of the HEPA (high-efficiency particulate air) vacuum cleaner filter. At that time, it was only used in a few devices to remove fine particles (down to 0.3 µm in diameter) present in the carpet or on the floor. Today, HEPA vacuum cleaners for the home are much more common. Basically, the HEPA is designed to capture even the fine particles disturbed by the vacuum cleaner. Historically, the HEPA filters were proven to be very effective in capturing radiation released by industrial and other processes. However, the most important feature of any good vacuum cleaner is the engineering. It must be designed to capture the particles as they are released from the surface with high efficiency, which requires a closed collection system and high capture velocity that is produced by the vacuum cleaner motor. Thus, in the context of WTC dust the use of a fan or an air blower to disturb the particles on surfaces in a room contaminated with WTC dust would be more than adequate to determine if the dust was resuspendable by a common residential vacuum cleaner.

I agreed with the EPA. Either form of testing was valuable in proving the presence or absence of WTC dust in the home, especially for those homes with more than just a trace of visible dust on the floor. Finally, the logical use of multiple repeats of testing and cleaning before giving a stamp of approval or certificate of removal of the WTC dust was excellent. By the way, I would have selected aggressive air sampling for my home. Very inconvenient, however, since you had to relocate during the testing. The results would give me or any home owner or renter a true picture of the long-term situation for resuspendability of the dust.

The facts about the homes cleaned up by EPA were that multiple cleanings were necessary to remove the WTC dust, and it could take up to four cleanings to remove one of the target chemicals of concern, asbestos or lead. The best part about this approach for WTC dust cleaning was that EPA used not only air sampling results but dust on surface sampling to determine how clean the residence was before finishing the job. Looking back on the issue, it was probably the best plan since at that time there was still plenty of dust from the collapse and disintegration of the structures present that would be easy to recognize in most residences. I still believe that few, if any, groups could have done as well as EPA in such a short period of time. You can read the EPA report on the program for more details (referenced in the bibliography).

When the Test and Clean program was over, however, lingering questions remained. The three greatest concerns to the public were as follows: Why didn't the EPA clean up whole buildings just individual units? Why didn't the EPA limit the activities to below Canal Street? What was the condition of some of the HVAC systems in the buildings? Other issues included these: Why didn't the EPA go into occupational settings? Why didn't the EPA have the ability to go into people's homes without permission? The details on the latter points are beyond the scope of my expertise to explain in detail, but it was the province of OSHA to go into workplaces, and it would be very difficult to just go into people's home without permission. The latter would be viewed by many as an invasion of privacy. Many lawsuits would have resulted, and nothing would have been accomplished.

The other three topics leave room for comment concerning the WTC dust. I think that in the beginning of the aftermath the "whole unit" approach to cleaning would have worked. However, considering the lack of a cleanup plan at the beginning of the restoration phase of the area around Ground Zero, and the fact that people had already started to clean up their units, it would have been impossible. As I reviewed the situation in 2008, each unit would present different problems in WTC dust loading and character, and the degree of cleanup required for each apartment or condo; thus, whole-unit cleanup of WTC dust still would have been impractical. Historically, do people clean up whole units at the same time after the effects of natural or man-made disasters? Or do they complete the cleanup of individual units? Probably more the latter than the former. In the end, downtown Manhattan was primarily cleaned up by professionals either privately or by the EPA Test and Clean program.

On the issue of the extent of the cleanup, the EPA ended its Test and Clean program at Canal Street and did not cross over into Brooklyn. I had a major disagreement with EPA on stopping at the end of Canal Street. However, at that time, we still did not have a WTC signature, and the farther one's residence was removed from Ground Zero, the more likely the chemicals of concern used by EPA—asbestos and lead and dioxins—would have been mixed with background levels of the same materials. As a result, lower loadings of WTC dust on surfaces in the residence would have been indistinguishable from other dusts. However, our WTC dust sample collection and anecdotal information did suggest that areas north of Canal Street were of concern. It may have been prudent to do a visual inspection of the residences of concerned individuals to determine whether visible WTC dust was present, to provide a degree of closure for some individuals, but as you can readily appreciate, that is now part of history and can no longer be done.

In terms of the background dust study done by EPA, the design included testing twenty-five residences and nine common areas located around Manhattan that were known *not* to have been directly impacted by the WTC dust plume. The EPA report became available in April 2003, which was well beyond the time when it would have been useful for defining the cleanup values for the Test and Clean program, but it did set the stage for any future terrorist events or natural building collapses. The study eventually was able to be used in further discussions about cleanup; however, the potential future use of the data in planning for disasters would be an important step forward in understanding the nature of residential dust. I wish we had that data early in November 2001. This good study of background dust became lost in the noise that was rising within downtown Manhattan in 2003.

On about the third anniversary of the attack, the American Chemical Society held its meeting at the Javits Center in Manhattan and included a major symposium on the WTC aftermath. The science was well received by a relatively modest crowd. There was a mixture of presentations on the WTC dust and the air quality around Ground Zero. I had nothing new to say, just offered some additional insights that were essential in describing the time line of the events associated with the WTC aerosol during the aftermath. I became more critical during the press conference and question-and-answer period because again there were more and more claims about the worst air in NYC ever during exposure period 4. This reaction was over the top. A person has only to look at the historical record for NYC and other urban areas in the United States prior to the signing of the Clean Air Act in 1970, and it is very clear that the air quality was worse during that period of time, not after the attack.

I vividly remember that as a kid living in Passaic, New Jersey, one of my jobs was to go outside in the morning during the heating season and wipe the coal dust particles off my father's car. They fell out of the atmosphere during the evening when the coal-burning heating systems were stoked to keep us warm, but other coal combustion emissions remained in the air. This type of "heating season air pollution" led to a grayish smog during periods of low winds or an inversion. My personal experience with heating season–type smog in NYC was during the Thanksgiving episode of 1966. It lasted a few days and was blamed for the deaths of over 150 people. We were traveling to Poughkeepsie, New York, to visit my mother's family for the holiday. We passed through the smog on our way up and back. What was most interesting was that after we left the inversion that had settled on New York City, the smog disappeared. Why do I remember this so well? It was the last Thanksgiving I spent with my grandfather. He had contracted leukemia the year before and was a test patient at Brookhaven National Laboratory for chemotherapy. I believe to this day that he would not have lived that extra year if he had not been enrolled in their experimental program.

I also remember working with the coal boiler fire in our home to keep clinkers, an uncombustible residue that has been fused into a lump after burning coal, from building up in the furnace. Much to my mother's dismay, one of my wintertime "fun" activities was to watch the coal come down the coal shoot and then slide from the top to the bottom of the new pile! Clearly, my clothes got very dirty.

At the ACS meeting, I had to disagree with Dr. Cahill's conclusion on the air quality issues, since he called the plume in October 2001 a "toxic soup." By October 2001, the plume had significantly decreased in intensity, and the air quality was affected by only brief excursions above routine pollution values in NYC. It would not be fair to say that Dr. Cahill was totally wrong since the composition was still not back to pre–September 11 in character, but the reductions in levels that had occurred since the 11th all pointed to improved air quality. I felt that the WTC air quality issue did not gain any substantially new information at the meeting. Some new chemicals were mentioned by Dr. Cahill, such as fused iron spheres, and they added to a long laundry list of compounds released during the collapse and fires. There was a summary of EPA sampling and analysis data collected during exposure period 4, and Lung Chi talked about the high pH. In my mind, it was a wrap-up of the scientific issues related to the WTC dust and NYC air quality during the months after the events of 9/11. However, I did make the point, which was alluded to by others, that society will never know all that we should have or could have about the composition of the WTC dust and the gases that were released during the first hours and days (exposure periods 1 and 2).

After the meeting, Christine Todd Whitman, the former EPA administrator, wrote an op-ed piece in the *Star-Ledger* on the WTC dust and air quality issues (September 14, 2003). She mentioned the presentations of Dr. Cahill and myself at the ACS meeting, indicating that we supported the accumulating results from EPA. In her statement, titled "The EPA Was Right," she said, "The latest air quality report confirms what the Environmental Protection Agency said. Paul Lioy (EOHSI, Piscataway) and Thomas Cahill (U. California) found that the plume rose quickly and posed little threat to the people on the streets. Their study joins many reaching the same conclusion" (Whitman 2003). On this point I would agree, since she was focusing on the nonoccupational outdoor exposures that occurred beyond the first two days. I would have reminded her, however, that the NIEHS team had first mentioned the significant reductions in outdoor air pollution by the end of 2001. I have never had a detailed discussion with the administrator/New Jersey governor, although I know many on her staff in New Jersey and in the EPA. We could have an interesting conversation on WTC dust and the aftermath.

I know individuals today who are still analyzing the aged samples of WTC dust. For example, in 2007, an editor from *Esquire*, Eric Gillin, who had stored a backpack he wore during the 9/11 collapse, had it analyzed for WTC dust by Dr. Cahill. Gillin described this experience in the April 2007 issue of the magazine. It was an interesting human interest article that added to the story of the WTC dust. A point made by Dr. Cahill was that the material "from five years ago was light and fluffy and had not gotten humid or moist. It was the closest we're going to get to the stuff that was in the air for a person who was running in the street." Sounds like the characteristics of our three settled WTC dust samples, which indicated that this was truly an analysis of WTC dust. I am sure that many who had been covered with the WTC dust that day were given a vivid reminder by the publication of the *Esquire* article and now by this book. By the way, for those interested in the *Esquire* article, it was published in the April 2007 issue with the actress Hilary Swank in the cover.

⓫

WTC DUST STICKS LIKE GLUE

In the summer of 2003, I had concluded that the WTC dust efforts were slowly resolving themselves. Reports were coming out, scientific manuscripts were being published, and the Test and Clean program was over. Thus, with the exception of the issues surrounding the long-term health of individuals who were without respiratory protection and breathed the WTC dust and combustion gases during the first few days, the story should have come to an end.

In September 2003, Jeanie and I went to the International Society of Exposure Science meeting in Stresa, Italy. There I had an invited plenary presentation on homeland security and issues related to human contact with highly toxic material. This was the first of a number of presentations that I made through 2008 that began to draw upon the WTC experiences to identify important gaps in our understanding about how to respond to disasters and the need to use the principles of exposure science. I had become very intrigued by the way in which we respond inconsistently to events that appeared to reveal carelessness, unawareness (now known as situational awareness), lack of efficiency, and the lack of coherent strategies to characterize exposure. Although most people have some grasp of the meaning of a hazard, which is the basic toxic property of a material, very few have a clear understanding about the importance of *contact with a material* and characteristics of *subsequent exposures*. It is my firm belief that development of exposure concepts, and the use of exposure science tools, is required by a broad spectrum of individuals to improve our understanding of real human risks to environmental and occupational diseases. Overall, educating the public on this issue has been a very slow process, and the actual WTC dust and experiences obtained during the aftermath require more activity to reduce our ignorance about exposure. Personally, this led to my participation in a number of research projects, the TOPOFF (Top Official) exercises, and other homeland security–related activities. However, I will leave this discussion alone for a while because the WTC dust came roaring back into my life after we returned home from Italy.

During the late fall of 2003, however, we entered a new phase of the WTC dust story. The formal activity became known as the WTC Technical Panel. I was invited to serve and asked to be vice chair. At first this appeared to be an honor, and given the goals and objectives, my time spent on the panel should have been worthwhile. The formal panel announcement was released Monday, March 1, 2004, and the chair was the assistant administrator for research at the EPA, Dr. Paul Gilman, who was serious about achieving the charges that were placed before us by the Council of Environmental Quality. These read as follows:

WTC Air Quality Expert Technical Review Panel
Purpose: To provide greater collaboration in ongoing efforts
                             to monitor the situation for New York workers and
                             residents, we will convene an expert technical review
                             panel to help guide the agencies' use of the avail-
                             able exposure and health surveillance databases and
                             registries to characterize any remaining exposures
                             and risks, identify unmet public health needs, and
                             recommend any steps to further minimize the risks
                             associated with the aftermath of the World Trade
                             Center attacks.
Representation: EPA will organize and lead this group of experts,
                             with representation from the federal agencies di-
                             rectly involved in the air quality response and moni-
                             toring, the New York City Departments of Health
                             and Environmental Protection, and outside experts.

Charge: The panel will:

By April 27, 2004, 1. Review post-cleaning verification sampling in the residential areas included in EPA's Indoor Air Cleanup to verify recontamination not has occurred from central heating and air conditioning systems;

2. Review the peer reviewed World Trade Center Indoor Air Assessment and Selection of Contaminants of Concern and Setting Health-Based Benchmarks, which concluded asbestos was an appropriate surrogate in determining risk for other contaminants.

By October 27, 2005:

1. Identify areas where the health registry could be enhanced to allow better tracking of post-exposure risks by workers and residents.

2. Review and synthesize the ongoing work by the federal, state and local governments and private entities to determine the characteristics of the WTC plume and where it was dispersed, including the geographic extent of EPA and other entities' monitoring and testing, and recommend any additional evaluations for consideration by EPA and other public agencies.

Operating Principles: The technical expert panel will be guided by the following principles—

- All meetings of the panel will be held in the New York metropolitan area, open to the public and announced in advance.

- The panel will be chaired by an EPA representative, and a non-governmental panelist will serve as vice-chair. The chair will consult with the vice-chair in the preparation of all panel documents, including meeting agendas and recommendations.

- Individual members of the expert panel will represent themselves and not their respective institutions.

- Panel members will be asked to disclose possible conflicts of interest.

- Panel members will be asked for their individual comments and recommendations regarding the charge. There will be no attempt to reach consensus or to develop group recommendations within the panel which might otherwise stifle individual opinions.

- The panel members may provide comments and recommendations in writing and verbally at public meetings. They will not make decisions or develop group positions. The Chair, in consultation with the panel, will summarize and submit comments and recommendations pertaining to each part of the charge to the EPA Administrator and other appropriate state and local organizations. However, it will be the responsibility of EPA and other appropriate state and local

organizations to formulate responses to the recommendations that
represent the overall advice and recommendations of the panel.

- Ex-officio panel members, including the community liaison, will
  receive all panel materials in advance of meetings, will be able to
  submit comments and recommendations in writing and, will be
  able to briefly summarize their comments and recommendations
  verbally at appropriate times during panel meetings.
- Detailed minutes of each panel meeting will be kept.
- Panel documents will be made available on the Web (to the extent
  technically possible) for contemporaneous public inspection. Otherwise, documents will be made available through other means.
- Composition of the expert panel may change as required for the
  different phases of the charge.
- EPA will brief interested members of the New York, New Jersey
  and Connecticut delegations and responsible Congressional committees bimonthly or as requested throughout this process. (see
  http://www.epa.gov/wtc/panel/)

The final list of the panel members is in appendix B. The members were
from a wide range of disciplines and affiliations, and membership was
completely voluntary. I knew many but was introduced to a number of
very interesting people from different backgrounds. They also had dif-
fering perspectives on the WTC dust, health effects, and cleanup issues.
I was looking forward to the challenge.

The list shared here describes lofty goals that we should have been
able to achieve in a timely manner. Unfortunately, there was one main
obstacle: as a group, we were not allowed to vote or obtain consensus or
make a final decision on any issue. Basically, as underlined in the charge,
that meant we could only provide our individual comments, which can
never be an optimal way to complete a job. Imagine trying to build a
bridge and everyone had input to the design, but no one was allowed to
make a decision or achieve closure on what would be the final design. Do
you think the bridge would ever be built, built well, or built in a timely
manner? Also, think about serving on a local committee or council. What
would be its value if you could not write a consensus report or present
substantive recommendations to those who had appointed you?

Because the panel could not make recommendations on any topic,
the panel activities seemed doomed from day one. I should have walked

away and not volunteered to serve on such a panel. However, I felt a sense of obligation to the country so I stayed on. The panel also had its introduction on March 1, 2004, which was a meet-and-greet with Senator Hillary Clinton, who had fostered the concept of a technical panel. Our first meeting was held on March 31, 2004. We also had a community liaison, and I was hoping that we could work to meet the charge questions quickly since it had been over two years since the attack.

I must mention that the WTC Technical Panel was a completely voluntary effort. In fact, the nongovernment employees were not even given local travel reimbursement to attend each meeting! The panel started with a lot of enthusiasm about tackling the charges we were given to address. During the time that we were being formed and organized, the EPA scientists from Region II were simultaneously trying to develop next steps for a follow-up program regarding the WTC dust cleanup and focus on "unfinished business." I must state clearly that the scientific discussions designed to help the local community were eventually overshadowed by the community's agenda, which was not necessarily science oriented. Thus, the initial traction made by the panel during its first meeting on the EPA draft plan quickly slowed down, and I began to realize that indeed the panel had little power to get things done.

I do have to commend Paul Gilman for working toward a scientific solution on how to best achieve a solid plan toward a second cleanup. Unfortunately, it was clear that the community and the panel were going in different directions. Many on the panel (although there were no votes or consensus) felt that a WTC signature (fingerprint) as well as other technical issues, including a determination of the extent of WTC dust in the area, were essential for a next phase of sampling and then cleaning. However, it became apparent that the community wanted to talk about other issues, and as soon as they learned about the availability of CBPR (community-based participatory research from the EPA), they wanted a grant. Conflicting goals of the community became a distraction to the panel, slowing down the overall discussions on the plan to proceed with the development of a WTC signature and address other scientific issues.

I must note my total disappointment with the panel process and the lack of appreciation of the committee by a few members of the community. It was my feeling that the CBPR was being pursued to address many issues beyond the scope of our panel's charge. I have no firm evidence, but during each meeting it became apparent that science was not a major concern, and in fact each meeting was dominated by repetitive and long-winded public comment periods. We started with one comment period and eventually this evolved into two comment periods. The meetings began to feel more and more like a public meeting on the general issue of WTC and can we go back to September 10, 2001. This change in direction should not have happened, and many regretted the change since the comments never ended and the topics continued to expand. Furthermore, it led to more frustration among the community and difficulty in keeping all focused on the charge.

Problems did not end there. A small number of community members became, at minimum, sarcastic or nasty, and over the course of the twenty months during which the panel existed, we were read poems; serenaded with songs, "a cappella" or with instrumental accompaniment; called names; and finally picketed. In all my years of dealing with community-based problems in inner cities or elsewhere, I never experienced such disdain for the people on a technical panel who were only there trying to help. We also had no power to address some of the issues of concern to the community, which caused even more tension between the two groups. There should have been another setting or committee to address the wide-ranging issues that were troubling the community. Further, except for about $7 million available from FEMA, we had no ability to address the magnitude of the budget required for cleanup, or the possibility for requesting more resources. This was an overarching problem that was never resolved.

I also learned that I was a blue-shirted puppet of the administration. Still, I can understand the frustration of members of the general public, but some of their goals were unrealistic, and to paraphrase a nursery rhyme "we could not put Humpty Dumpty back together again." All we could do was understand the WTC dust better and provide guidance in designing any follow-up cleanups. Further, we wanted to assist in dealing with lingering health needs, but as you will see, that charge to the panel was never truly examined in detail.

It was never stated during our meetings, but it should have been said early on that *9/11 was not the first act of war perpetrated on a city, and though very tragic, it definitely was not the worst in the history of humankind.* It was, however, the end of innocence for these people in Lower Manhattan. The panel actually formed a subpanel headed by Greg Meeker of USGS, which included Mort Lippmann and myself to start drafting a plan for the WTC signature experiments. This was done after the initial presentation by the EPA, which did not address the issue in a proposed cleanup plan, when it was clear that the plan required a signature to firm up specific weaknesses. (This was done in spite of our inability to make a recommendation.) I have extracted part of our June 22, 2004, panel summary that is still on the EPA WTC website, to illustrate what the issues were and the time line for the WTC dust signature. It starts with a summary of the presentation made by Dr. Morton Lippmann at the meeting.

Mort defined what a WTC signature should be:

> A set of analyzable materials, elements or chemical compounds which individually or in combination provide adequate evidence of contamination from World Trade Center dust and/or combustion products at concentrations determined to be of significance. (U.S. Environmental Protection Agency 2004)

He explained that he and the subpanel members (Meeker and Lioy) were tasked to suggest guidelines for the development of the signature. Mort also presented a realistic time line (table 11.1) for the development of the signature and noted that these estimates may be delayed by a month.

The "doable" proposal was designed to find a signature (previously called a fingerprint) for "the presence" of significant levels of WTC dust, in the context of the contaminants of potential concern (COPCs) and any new materials found in analysis of the WTC samples. Paul Gilman indicated that "if a signature exists, then archived samples would be used to quantify the relationship of the signature to COPC, and then new sampling would be conducted to verify those relationships" (U.S. Environmental Protection Agency 2004).

The age of the archived samples, which were two and a half years old, was questioned, because the WTC dust may have changed over time. I noted that one of the goals of the signature study was to identify materials that would *degrade least* over time, which is a typical methodology for analyzing archived samples. The answer was yes, the samples would degrade over time. The good news was that any degradation of the WTC dust composition would actually help rule out candidate materials for a

Table 11.1. Tasks and Time Line for the Validation of a WTC Dust Signature for the WTC Technical Panel in 2004

Approximate Schedule	Task
June 2004	• Develop guidelines for participating research labs. • Establish sampling subgroup to identify methods, protocols, and analysis.
July 2004	• Compile list of archived samples and results.
August 2004	• Identify and arrange for participating laboratories. • Develop requirements for sampling.
Immediately after authorizing laboratories	• Begin analysis of archived samples.
1 to 2 months after authorizing laboratories	• Provide interim report on robustness of proposed signature components.
2 to 3 months after authorizing laboratories	• Begin collection and analysis of "typical" test samples.
3 to 4 months after authorizing laboratories	• Define detection limits for signature components. • Establish analysis flow chart.
Spring 2005	• Final report on success of WTC signature development. • Determine analytical benchmarks. • Peer review.

signature since these would probably be degraded in the residences and other places that would be considered for cleanup.

From the meeting summary, a panel member was concerned about "how . . . a signature would be different than sampling for all COPCs and fibers." Mort Lippmann noted "that the elemental X-ray analysis enables the identification of slag wool, cement, and gypsum at much higher sensitivities than the COPCs. Therefore, in some cases, the Chemicals of Potential Concern (CPCs) may not be analyzed [detectable], but the analysis can identify the signature."

One issue discussed many times was that the presence of the signature would not necessarily trigger a cleanup or a recleanup of a residence. To be of value to the local community, the signature needed to have an action level or threshold for cleanup based on the presence of enough WTC dust. During a panel meeting, one panel member, Mr. Newman, asked us to also consider the significance of finding no signature in a sampled space. We tabled that point, and in hindsight I think we should have addressed that issue from the beginning. It would have provided a basis for encouraging the development of a signature among the members of the community.

Slag wool had already been mentioned as a possible signature. Mort clearly stated at the meeting that, in his own judgment (no consensus, but I agreed), "the lack of the presence of the slag wool at this expected low detection limit would indicate that WTC contamination would be below a level of concern. Therefore, there needs to be a definition of the significance of identifying slag wool in a sample." This was an excellent answer that should have been understood by all. We were not trying to develop a signature that will define "how clean is clean." The work being conducted to develop a signature was meant to identify presence or absence of significant levels of WTC dust.

One panelist brought out the point that we should also consider the rate of false-positive results, since slag wool was used in other buildings. This was true, but the results from the background study, and the experimental design for the signature study, were supposed to address the issue. The WTC dust had been resurrected. I felt that scientifically we had achieved a milestone even though the panel's agenda was expanding and the meetings were deteriorating. Maybe we could get closer to identifying a signature for WTC that was better than a visual observation. I felt that the signature was important in implementing an effective revised cleanup program.

The EPA National Exposure Laboratory was part of the effort as well as the USGS. We were fortunate in having both these laboratories willing to participate, but communication failures seriously reduced the effectiveness of the signature program. This was in no small part due to the panel structure and its lack of any ability to achieve consensus.

During all the later meetings of the panel, we spent about half our time dealing with many issues that were beyond the scope of the panel and also with the public's concerns about the dust. I began to feel that substantive discussions on science and other technical issues were being lost, and our focus on the WTC dust was being questioned at every turn. In layman's terms, we were stuck with a public that wanted a total cleanup of southern Manhattan for chemicals of concern irrespective of the source. This was a utopian goal that was too far outside the panel's field of view or society's ability to achieve.

The meetings went on. They were frequent through the fall of 2004, and for the interested reader, summaries can be found at the EPA WTC website (http://www.epa.gov/wtc/panel/meetings.html). The bulleted lists on the first few pages of each summary represent the litany of comments made by the panel members on a variety of topics since there was no ability to reach consensus or general recommendations. So depending upon where you were sitting, there was smorgasbord information and at least one comment was going to fit your particular bias or point of view.

Some examples of truly pertinent panel-related topics were the extent of any new cleanup needed to go beyond southern Manhattan into Brooklyn and above Canal Street, and there was a need to consider cleaning HVAC systems.

One positive panel-related activity was that for the first time, an overview of all the work done by USGS, my team of collaborators, and others on the WTC dust would be done at the same time! Eventually, I also supplied stored portions of our two-and-a-half-year-old samples to Dr. Meeker for his signature experiments. I felt that this was reasonable because any WTC dust that remained in a residence and would be subjected to a cleanup would also be two and a half years old. Thus, the results would be directly comparable for all WTC dust samples. The one new twist that Greg was going to provide was dilution scheme to determine how little WTC dust was necessary to be in a total dust sample before a signature was lost—a great idea. Thus, you can envision an experiment where the WTC dust would be diluted with other dust, and the dilutions would become larger and larger until Greg could determine that you could no longer detect the WTC dust under the microscope. His reanalysis of our Market Street sample is shown in figure 11.1, and it was found similar to the results obtained from his samples, filling in some details on MMVF and gypsum.

By the July 2004 meeting, the EPA had begun to consider expansion of the study area for a second cleanup program. Things were beginning to look better for the science of the WTC dust, especially since the EPA was planning to validate the existence of a WTC dust signature. However, the community demands were increasing, and our panel could not address most. By the end of the July meeting, clearly half of the time was devoted to community comments and not the tasks of the panel, which made a difficult task even more difficult.

The October 2004 meeting of the WTC Technical Panel was a turning point in our activities. It was then that the WTC dust almost disappeared

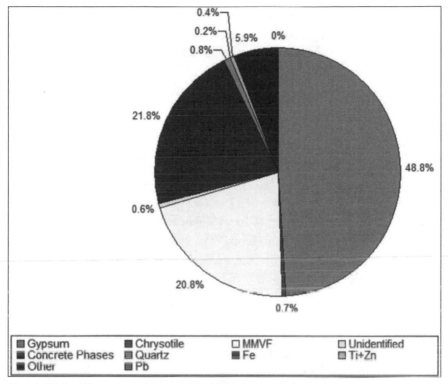

Figure 11.1. The composition of the WTC dust sample taken at Market Street by Weisel and Lioy, and as analyzed by G. Meeker. Reprinted with permission of Dr. Gregory Meeker, USGS, Denver, Colorado.

from the discussion. The community went on an offensive, presenting seven principles to the EPA. I was never sure what principles were being mentioned. It was more like seven demands of the EPA. These are found on the EPA WTC website. At this point, the community could care less about a signature; they wanted EPA to clean up southern Manhattan, as well as many other things.

At that time, though lost in the noise, was the fact that EPA's portion of the validation study actually had secured ten WTC dust samples and ten non-WTC dust samples for analysis. You can read the transcripts, but the big WTC dust question, which should have been tabled for a while, was, What do we do if there was no signature?

Another signature issue was brought up, that of a fire signature. An interesting concept, since we had tried to obtain a PAH signature for the dust. Our PAH signature was too noisy and did not have enough specificity, since many types of fires will occur periodically in any city. The question was whether we could extract a PAH signature for the plume and whether samples were available for such an analysis. A signature for the fires would have gone a long way to differentiating WTC dust from the plume of smoke.

We were also getting close to identifying a new building cleanup plan, but without the ability to reach a recommendation, the discussion continued past October. I returned my comments on the EPA plan by November 3, 2004, and we had an excellent conference call on November 16, 2004.

A major finding presented by Meeker at that time was that the MMVF contained over 90 percent slag wool. The first proposal for a signature! He also found a complicated signature based on thirteen components detected by another lab, but it would not be easy to duplicate or easy to implement effectively within the population of buildings in southern Manhattan.

In the meantime, Lung Chi finished his research on the size fractionation of the WTC dust collected by his staff at fifteen outdoor and eleven indoor locations, all close to Ground Zero. The work reaffirmed two major points from the 2002 paper that was discussed earlier. Namely, the size distribution of all samples was heavily weighted toward the size range above 10 µm in diameter; these were inspirable particles or the supercoarse fraction. The pH of the samples was primarily above 10, indicating alkalinity with the additional fact that the pH of the smaller particles was closer to neutral, or pH 7 (see table A.8 for more details). This was an important finding since it showed again that the World Trade Center cough was closely coupled with the nonregulated supercoarse particles. However, the neutrality of the smaller particles should not be misinterpreted to mean that there were no cement particles in the smaller size fraction. Work by Millette, published in 2002, found that cement particles could be found in the fine size range. The neutrality would be derived from the fact that the smaller particles would contain a higher proportion of the combustion products than the particles in the supercoarse particle size range. These organic particles would be neutral or more acidic, which would decrease the high pH that would be found in the cement particles.

The techniques used by Jim could not detect the organic components, which supported the decision we made to measure just about everything we could in the dust to get a full picture of the complexity of the WTC dust.

The next meeting of the panel continued to focus on too many issues. We did have a short discussion on the WTC signature validation. To my surprise, we learned that the EPA and not USGS was taking the lead on the signature work. This was a major change, and I felt it led to a series of missteps because too many managers became immersed in the process. In the end, it led to the signature moving far away from a cleanup signature to one that would define how clean is clean for risk analyses. I also felt that with the extreme tensions that existed between the EPA and the local community, a more neutral agency should have led the effort. This was another decision that we could not approve, and it would come back to haunt everyone at a later date. Again the story of the WTC dust turned another page.

At the November 2004 meeting, the community commented extensively on the EPA cleanup plan. If you read the transcript carefully, you can see that without the ability to make recommendations, the community questions could never be put to rest, which increased their frustration and illustrated the weakness of the panel. The community, as part of the CBPR, hired a set of experts (but, remember, we were not compensated for our time) to comment on the EPA plan. As I would have expected, the comments were extensive. Their bottom line basically was clean up lower Manhattan, and obviously do not wait for a signature. The WTC dust signature effort limped on, and the comments and demands continued, but any chance for providing solutions was slipping through our fingers.

The sudden resignation of Paul Gilman as the EPA assistant administrator of the Office of Research and Development (ORD) at the end of 2004 resulted in the panel beginning to fall apart in 2005. Furthermore, we were still not moving on to the panel's charge of examining "unmet health needs." The EPA appointed an interim chair, Mr. Tim Oppelt, who was also the acting assistant administrator of the ORD. After about two months of discussion and meetings, it seemed clear that Mr. Oppelt just wanted to finish. In some ways, I could not fault him; he had inherited a situation from which there was no simple solution or easy exit strategy. He called meetings much less frequently, which, based on the content of each meeting, was no loss. However, it was also clear that he just did not understand the signature-development process. Furthermore, the EPA in its efforts to validate the signature missed the point. It wanted to determine "how clean is clean," which was not our goal. Again, the original goal was, Can we find a signature that could validate the need for a cleanup of a residence?

A report by Greg Meeker said it all:

> If slag wool fibers are not found in settled dust above a predetermined critical level, it is unlikely that the COPC derived from the WTC could be present at significant levels in the samples. . . . [I]n difficult cases the size distribution data . . . might prove useful in distinguishing WTC materials from similar materials from other sources. (Meeker 2005)

Remember, the supercoarse particles were relatively unknown to everyone. It is a pity that the points in Greg's report were not considered by Mr. Oppelt before he convened his own panel, another expert EPA panel to review the signature results from EPA. It was supposed to "provide recommendations to our Expert Technical Panel." In the end, the Oppelt review panel, which was composed of many good professionals but uneducated to the purpose of a new WTC dust cleanup program, missed the point. They came back with a report that Greg, Mort, and I strenuously objected to upon reading and in conversation. Our comments on the signature review are found in appendix C, and again much more detail is still found on the EPA WTC website.

The bottom line was that the three of us strongly agreed that there was a signature. Unfortunately, in the end all we achieved by the second review was what I would describe as "paralysis by overanalysis." Our panel should have been allowed to complete the charges used to form the panel, and we should have been encouraged to complete our job effectively. This is something to think about prior to forming any committee.

On November 2, 2005, my last major interview on the WTC dust was included in a *New York Times* Metro section article by Anthony De-Palma. It was in response to the Oppelt panel report, and at that time we (Meeker, Lippmann, and I) were still formulating our response to the peer review panel formed by Mr. Oppelt. However, I did indicate that slag wool would be just fine as the signature considering the massive quantities that would be present in uncleaned spaces and the very odd size distribution (i.e., supercoarse particles).

As a bit of a digression, during the interview and in the article, Tony (DePalma) said that the *tone of my voice* changed when I discussed the WTC dust with him or in public settings, including during a lecture at Montclair State University, New Jersey, on October 6 earlier that month. Jeanie had made the same statement during other speaking engagements. I realized it was part of the change in me caused by the work we had done on the WTC dust and the impact the attack had on me and many others. I stated again that the WTC dust held everything we hold dear, and it does, even if in undetectable microscopic quantities The mention of the WTC dust did lead to other points of view. As printed in same the article, Catherine McVay Hughes, a resident and the WTC Technical Panel community liaison, stated that "[s]he cannot stand to be near it anymore and she keeps five air filters in her home even though it was cleaned up three years ago" (DePalma 2005).

The committee received a letter from Mr. Oppelt toward the end of November 2005. I had reviewed a draft but did not agree with the content. It was about the plans of EPA, and it announced the final meeting of the panel, December 13, 2005. To place the letter in the proper context, I will refer to the statement issued on November 29 by Senator Clinton just before our final meeting.

> While the plan includes modest improvements over the prior EPA indoor cleanup program, it fails to correct the major problems identified by EPA's Inspector General in 2003. For example, the plan does not include testing in north of Canal Street or in Brooklyn, in spite of the Inspector General's conclusion that the cleanup boundaries were not scientifically developed. In addition, the plan ignores many of the recommendations made by the World Trade Center Expert Technical Review Panel over the last 20 months. Finally, the EPA's proposal today to disband the Panel after the next meeting is unacceptable. The Panel has not even begun to meet its mandate to identify unmet public health needs and recommend any steps to further minimize the risks associated with the aftermath of the World Trade Center attacks. I will be fighting to ensure that the Panel completes this important task.

It was a pity that the panel could not make the recommendations suggested in the Clinton announcement. The conclusions made by EPA in the Oppelt letter included going ahead with a new Test and Clean program. Thus, the reason for the meeting was confusing at best.

The meeting held on December 13 was a disheartening affair for everyone. The EPA had made the decision to reject the signature and recommended a program similar to the Test and Clean program completed in 2002. For the community, it was a total rejection of their seven principles, and it was also a total rejection of the WTC Expert Technical Panel. The latter opinion was expressed in many different ways by individual panel members. After the meeting on the 13th, I stated to David Prezant in an e-mail, "I have been asked to consider a book on the WTC, and I was having a hard time on a title. I think I have one now: 'From Optimism to Disillusionment: My View of the Use of Science in WTC Aftermath.'" Obviously, the title changed over the course of time to the one on the cover of this book. There was too much interest in the story to let this chapter of the WTC dust dominate my thoughts.

For the first time in my professional life, I had seen the deliberations of a scientific committee/panel totally implode. The EPA management forgot the purpose of the panel, and in the end the process alienated both the panel members and the community. You can find the details on articles saved on the Web by entering "Oppelt and WTC." Oddly, by the end of the meeting, Mr. Oppelt changed his mind on the EPA pronouncement about the signature. He said that there should be more discussion about the signature. A comprehensive summary of the meeting on December 13 was published in *Chemical and Engineering News* titled "Terrorism's Legacy: EPA's Plans to Test Homes for 9/11 Contamination Sparks Anger among Advisers and New Yorkers" (published in 2006). In the article, Mr. Oppelt said that in response to the panel's criticisms the "EPA would reconsider the peer reviewer's comments on the use of slag wool as a signature for WTC contamination. We gave up too soon. We got it wrong."

He also said the Test and Clean program would not be expanded to other areas outside downtown Manhattan and that the panel was finished. That was a definite relief for me. This was a truly confusing ending to a very unusual day. The meeting was marked by picketing, people confronting the chair, and lots of name-calling. It was not a pretty picture. I was just happy it was over.

The *Chemical and Engineering News* article told the tale without many of the graphic images or much of the verbal commentary. There was a follow-up with EPA during the next few months, but in the end nothing changed, and the EPA started moving on with the second Test and Clean program. By mid-2008, only four hundred individuals had responded to the new program.

Obviously, the panel never did get to seriously discuss unmet health needs. That issue has lingered through today with federal testimony and hearings and some appropriations for those "injured and sick" during the aftermath. Nor did this panel ever issue a final report—a fitting anticlimax. However, I never got the opportunity to complete one of the charges, reporting to the congressional delegation. This could have been an interesting day, but we all moved on and the panel just died.

After the committee was disbanded, I thought we were finally done with the WTC dust. In February 2007, Jason and I went to the Yogi Berra Museum for the first showing of a stored reel of the October 8, 1956, World Series Perfect game, which I had attended with my father. It was a good retro moment that was written about by Allen Barra in the *Wall Street Journal* and then in his book: *Yogi: The Eternal Yankee*. However, during the period from early 2006 through 2007, our signature subgroup continued to work together on the WTC dust, and we finished a scientific paper on the WTC signature in 2008. The manuscript was published in 2009; Greg, Mort, and I were among the authors (see Lowers et al. 2008 in the bibliography). It reviewed studies on WTC dust and determined which components of the bulk WTC dust were above typical background dust levels in New York City. We focused on the less than 150 µm size fraction. Based on all the previous work, it had large quantities of gypsum, phases compatible with crushed concrete, man-made vitreous fibers, silica, lead, chrysotile asbestos, and other materials.

Slag wool was reaffirmed as the "most common WTC MMVF," while soda-lime glass and rock wool were minor to trace constituents. Most background samples also contained gypsum, phases compatible with concrete, and MMVF. However, the proportions of MMVF in background samples were typically unlike those which characterized the bulk of the WTC dust. Furthermore, the results showed that slag wool

could be used as a signature marker to identify areas that contain potential residual WTC dust contamination.

We were vindicated, but too late, and we hope that the analysis can be
of value in addressing what is important to do if another event similar to
the collapse and fires at the WTC happen again. I leave it to you to figure out how the above could have been prevented, ideas are welcome.
But to whom should they be sent? As with the Port Authority report,
who will listen as we approach the tenth anniversary?

A POSTSCRIPT

Two EPA scientists sent in a letter to the editor about our article. Instead of focusing on the science, it dredged up the panel and illustrated
some of the EPA's continued lack of understanding about what the signature was all about. The authors did provide some additional lessons
learned, which I basically agreed with. To summarize, more work has to
be done to develop an effective indicator; more understanding of background levels of dust should be achieved; and, finally, tight lab controls
were required for analysis. My question at this juncture is, Why didn't
the EPA do this over the past five years? Why was this recommendation
made by EPA scientists in 2008–2009?

EXPOSURE SCIENCE IN FUTURE CATASTROPHES

Since September 11, sessions have been held at many environmental and public meetings through 2007 on WTC dust, air quality, and WMDs. In 2008, however, we started to forget, and even at meetings that took on the issue of biological, chemical, and physical agents of mass destruction, attendance was decreasing. This was very disappointing since there is still too much unfinished business, especially on exposure issues. Now there are specialized meetings, many dealing with topics related to the Department of Homeland Security, but these have a narrower focus and are intended for more specialized groups of people.

Most desktop exercises do not ever deal with realistic exposure situations that can occur during and after events. Clearly, multidisciplinary activities are necessary, but without strong federal intervention, this is difficult to achieve among professionals. There is a definite need to fund research in exposure science and other programs for the field to survive in a very competitive world that includes homeland security issues. One positive note is that in New Jersey, Richard Canas, director of the Office of Homeland Security and Preparedness, initiated a preparedness college that has the goal of increasing the interaction among state professionals and the academic world. This type of effort, in other forms, has started elsewhere, and I hope it can bridge the typical disciplinary "stovepipes" and make us safer during catastrophes.

During the time when the WTC Expert Technical Panel was meeting, I was involved with two other very interesting activities that solidified my interest in homeland security issues and the need to include exposure science. These activities also led to the identification of gaps in how we respond to events from the standpoint of how people come into contact with, or can be prevented from coming in contact with, highly toxic materials, including weapons of mass destruction—namely, the types of activities and human behaviors that could occur during an event.

The first was the Urban Dispersion Program (UDP) Madison Square Garden Experiment, and the second was the TOPOFF 3. I know these are different from issues directly related to WTC dust, but they will help me explain what we need to know and understand to move forward into the future in the event of a catastrophe. I will frame them for you, and then connect the dots to the WTC dust.

THE UDP

The UDP was designed as a four-year study to determine how air flows in a city environment. It started in 2004 and ended in 2008. The results were to be used to improve and validate computer models that mimic the atmospheric movement of pollutants within cities and around, into, and within building interiors. Another objective was to improve the network of wind stations in and around New York City. Gary Foley of the EPA and I felt it could be much more.

The first of the field experiments took place in March, 2005, in the vicinity of Madison Square Garden. The second field experiment took place August 6–26, 2005, in Midtown Manhattan. I was involved with both studies along with Dan Vallero. We felt that the study we were going to complete would enhance NYC's ability to respond to many kinds of emergencies. It was this objective that we focused our attention on and for which we wanted to make a difference.

The movement of pollutants in the city was actually simulated in the study by releasing safe gases, called inert perfluorocarbons, at very low concentrations at specific locations in Manhattan. These tracers are safely used internationally by scientists and industry, and are easily detected down to very low concentrations: in parts per quadrillion, or one molecule in every 1×10^{15} molecules. The first day of the research program was a media event that had TV coverage from all over the world. I remember a TV crew from Germany intensely taking video of the tracer gas cylinder. This was sort of odd since the gas was colorless, but that's TV.

The study involved many organizations, including the city of New York, federal agencies, national laboratories, and universities. Gary Foley, who after 9/11 felt that this sort of experiment was a necessary addition for planning responses to future disasters, and I developed what has now come to be called a "prospective exposure study." The study also involved Drs. Panos Georgopoulos and Sastry Isukapalli of EOHSI, and it included two emergency response persons from EPA Region II, Drs. Eric Mosier and Jim Daloia. Eric and Jim made sure we did not lose sight of the need to address the emergency response issues in the design.

Our study can be explained as follows. The release of gases or particles that can occur in many accidental or deliberate events can have the potential to impact the health of first responders and the general public. To improve our understanding about how and where people might be exposed after a release, we conducted a prospective human exposure study in the atmosphere of NYC. The goal was to simulate a person's exposure to toxic gases by seeing how movements and activities immediately following a release may affect the intensity of short-term contact with an acutely toxic agent.

The results were better than I expected and provided information that could be used to predict a person's potential contact and then personal exposure to gases and fine particles in a complex urban setting. It was also to be used to reevaluate emergency response entrance and exit strategies, and community-wide stay-in-place versus evacuation strategies. (The latter is something that still remains to be finished today!)

The UDP study was a unique opportunity since we were able to replicate realistic release situations and exposures after a toxic release using safe gases. In combination with the safe gas emissions, we developed prescribed activity patterns that would be followed by members of our team and that may actually occur during or after an event. Sort of like the script for a play or a made-for-TV docudrama. You can easily see the connection with September 11, 2001, and the anthrax terrorism. We were trying to see the relationship between a person in harm's way and contact with toxicants that would be released at that time and place. Such information, if used properly, can improve evacuation, stay-in-place, and rescue strategies. Obviously, this was not an actual event. In contrast, in most "desktop" homeland security exercises that simulate a release, the victims are already defined, and the organizers provide a best guess of the numbers without much basis in reality about the contact rate and intensity exposure to the toxic agent. They make a lot of assumptions since it is rare that actual monitoring is capable of being taken during the height of an accident or other sort of event. Again, remember no one could be at Ground Zero during or just after the attack on the WTC to measure the exposures, and as noted previously, it would have been hard, if not impossible, to find any appropriate instrumentation.

To measure people's exposures to the "safe gases" released into the atmosphere in NYC, volunteer emergency responders wore simple but effective, pen-sized, personal exposure monitors that had been perfected by Brookhaven National Laboratories (BNL); these were provided to us by our collaborators from BNL, John Heiser and Paul Kalb. These monitors absorbed the "safe gases" over two- or ten-minute intervals of time during the completion of each prescribed activity. At the same time, other scientists tracked and measured the nontoxic tracer gases around the same area but with stationary detectors that had been placed on buildings or other objects. So we monitored both the air and the people during and after the time of the release. As a result, we could look at a timed release that would mimic how the safe gas would spread in Manhattan.

Looking backward in time to 9/11, we did not have a set of protocols and instrumentation available to measure anything during the WTC attack. This study provided an opportunity to understand some of the issues and others that would be associated with exposure to a highly toxic gas, the health consequences of an event, and what could be done to deploy simple exposure monitors during an event.

The study included two experiments with identical sets of subject activities, completed by the emergency response workers of the EPA. They collected the perfluorocarbons tracer samples that were designed to simulate exposure to a real gas when the responders were close to the source of the release, and when they moved about the surrounding neighborhood areas (up to seven blocks away). So, in other words, we measured exposures near the toxic release and in the surrounding neighborhood. We repeated these experiments with identical release strengths and along the same activity paths and patterns. The repeat experiment allowed us to study exposures at a different time of day and with some variation in the weather.

Having the EPA emergency responders as part of the team was important. First, they made sure the design was realistic, and second, they were the actors in the simulation. Being participants allowed them to actively participate and then make recommendations about how to use the data to help responders and the public. In most real situations, there is no time to design strategies you need to understand the potential situation as best you can. This was the goal of the prospective exposure experiment.

The Madison Square Garden experiment found that the highest exposures to the inert gases resulted from the activities taking place on the ground in the first few minutes following the release, but pockets of high levels could linger because of weather conditions and building locations and size. Furthermore, the exposures could not be predicted by simple models because local variations in size and spacing of buildings caused changes in wind flow among the buildings and locations farther away. The inert gases also were found in locations and at concentrations not anticipated from the modeling conducted prior to the study.

Finally, the Madison Square Garden experiment demonstrated that emergency responders and the general public need to be educated about the relationship between release location, the built environment, and the influence of meteorology to minimize contact with toxic materials. This meant that one has to determine whether staying in place versus evacuating is the right thing to do. In the case of a fire, just leave; in the case of a toxic gas, particles (e.g., anthrax spores or WTC dust), or radiation, you may walk right into a pocket of high levels of gases or particles. In a real situation, this could be deadly. For emergency responders, the results could be applied to establishing realistic but not perfect exclusion zones. These are areas that must not entered by anyone without significant respiratory protection.

I was very pleased with our work and the initial response to what we found by the emergency response community. It remains to be seen what the long-term benefits of our work are. The second phase of the work, the translation of our work to changes in community emergency response strategies, was not funded through the end of 2008. It seemed to get hung up in EPA bureaucracy, lack of interest, or a desire to drift back to familiar issues. It is a pity that this research and broader applications of the UDP results were not immediately followed up and published in the common literature.

The consequences to this lack of interest are unknown, but some positive news surfaced in 2009, when we did begin to get some funding to complete the second phase of the analysis of the UDP exposure data on the prospective exposure study. This occurred because of the persistence of my faculty, Jim Daloia (EPA) and Joe Picciano (formerly of FEMA and now of the NJ-OHSP).

Lessons from UDP

The results of the UDP experiments reaffirmed the weaknesses in the types of procedures that led to the confusion about exposures during the WTC aftermath. You do not have much time to respond, and confusion and chaos cannot easily be anticipated. Training needs situational awareness, but it must also be based on realistic problems and challenges. I think we get too comfortable, as citizens or as professionals. This leads to misplaced arrogance or confidence about our capabilities, and leads to failure.

I have great respect for those who risk their lives to save lives either as uniformed personnel or members of the military. That is why we did the experiment. To me, the best training for exposure reduction in the face of weapons seems to be coming from the military, which have simulators of all types and varieties. In the public sector, there needs to be an upgrade to situational awareness for high-consequence (acute toxic exposures) but low-frequency (a chemical tank car rupture) situations. In any field, you need to devise training tools for the public and professional with methods for reinforcing important concepts and using tools in the field again and again. The remarkable lifesaving part of the attack on the towers was the exit strategy. The employees had training and more training, which I and many others believed saved thousands of lives. But it was the experience of the 1993 attack on the WTC that helped harden systems and lead to a workable exit strategy.

In contrast, how many of us know how to get out of our homes in the case of a fire or where the fire extinguisher is located? Do you have evacuation or stay-in-place plans for a major catastrophe? Tools are available to fill this gap. Similarly, do the emergency response systems have a realistic understanding of what community or population exposures will be like in the case of a chemical or other attack? We do have the excellent work of the National Academy of Sciences, which has developed acute exposure guidelines (AEGLs) for many highly toxic chemicals. But who uses the AEGLs, and how do they get used? Do we have the appropriate instrumentation to measure levels in a real situation to compare with the guidelines? This can improve our ability to reduce the number of people who may die or get seriously injured.

We have learned many lessons about how to address the impact and aftermath of hurricanes. But there you have much more time to ready the systems and get people out of the anticipated harm's way. Without training activities that include the chaos caused during and immediately after an acute catastrophic event, it is hard to say whether a uniformed officer will even know where and when to enter or to back away until personal protection arrives. How many police officers have a fitted respirator in the front seat of their vehicles? We should use common sense to avoid a repeat of 9/11 where, in addition to the approximately three thousand people who died, almost six thousand rescue workers were injured by inhalation of the WTC dust and gases.

For you and your family, and me and my family, we all must have a catastrophe checklist and a home survival kit. Information on the contents of a kit or actual checklists are available from many reputable sources located on the Internet. These lists and kits should be reviewed each year, and you should have both an evacuation or stay-in-place plan depending on the type of event. There are even YouTube videos for emergencies such as tornados. Furthermore, we must devise an educational opportunity for the public to understand the nature of catastrophic events to be able to define what to do. Clearly, a family-based strategy for tornados needs to be different than a strategy for preventing exposures to your family after a chlorine tank car ruptures.

TOPOFF 3

The second Homeland Security activity I engaged in during the spring of 2005, which took place only about three weeks after the UDP experiments, was called Top Officials 3 (TOPOFF 3). At the time, it was touted, to paraphrase a press release, "as the most comprehensive terrorism response exercise ever conducted in the United States."

The event was sponsored by the U.S. Department of Homeland Security (DHS), Office of State and Local Government Coordination and Preparedness (SLGCP). TOPOFF 3 was the third exercise in a series that is a congressionally mandated exercise program. The exercise is designed to strengthen the nation's capacity to prevent, protect against, respond to, and recover from terrorist attacks involving weapons of mass destruction. Joining the Department of Homeland Security and other federal agencies were the states of Connecticut and New Jersey for this weeklong exercise.

The TOPOFF 3 took place from April 4 to April 8, 2005, and was two years in planning. The exercise involved more than participants representing more than two hundred federal, state, local, tribal, private sector, and international agencies and organizations, as well as volunteer groups. It was an important exercise. In Union and Middlesex Counties in New Jersey, the event simulated a biological incident that was caused by the release of pneumonic plague, a potent biological weapon. Obviously, a real weapon was not used, yet the response was mounted as if a release had occurred at a local university.

My initial reaction to the design of TOPOFF 3 was disappointment since from the view of exposure science, it lacked a realistic simulation of the release and contact by victims. From the standpoint of correctly defining personal exposure, the exercise was severely compromised. There was no definition of the chaos that would ensue after the event and during the time in which the symptoms began to appear within the community and beyond. During the exercise, participants were informed almost immediately about the nature of the attack and who the victims were. For plague, this approach was very unrealistic since symptoms do not occur immediately. This would be in contrast to a toxic chemical release. For a highly toxic chemical exposure, one can easily imagine that the immediate consequences of the release of chlorine from a ruptured tank car would be many sick or dying victims.

For plague, data on the exposure and number of victims would have been delayed for days, and you would not have known who was truly exposed and to what, especially the emergency responders who were called in to stop a vehicle in a parking lot. If the exposures had been designed properly, there would have been a heightened sense of drama and fear, and anxiety about what was released, and emergency responders also should have become ill. As is the case of many exercises, the people running the exercise designated victims and started treatment very soon after the release, which is not realistic. It would take days before people would have started to feel symptoms and start to go to hospitals. Unfortunately, the organizers of TOPOFF 3 had to compress the time between the release and onset of illness so much, and they did not tell most participants or the public that this was the case. Thus, to consider that the staged response would be viewed by many as well organized and efficient representation of exposure during an actual real incident could not be farther from the truth. The chaos that would occur in the first few days postincident was underrepresented, at best; nonexistent, at worst. During 9/11, we also did not have all the facts, which should have been the case in this exercise. Clearly, there were no weapons of mass destruction during 9/11, but more careful consideration of the potential acute exposures to the dust should have demanded a stronger implementation plan from health agencies during the first twenty-four to seventy-two hours postattack.

Numerous federal departments and agencies actively participated in TOPOFF3, and this activity was able to validate parts of the National Response Plan and exercise protocols of the National Incident Management System. Unfortunately, the script for exposure and acute response within the period of chaos was a failure.

As part of the New Jersey University Consortium for Homeland Security Research, I was asked to volunteer as a reviewer of specific activities, along with about twelve other people from the consortium, including Jeanie because of her expertise in microbiology. The outcome of our participation was again more education about the response to an event, and a total lack of understanding of time delays between exposure and health response, confusion, human behavior, and chaos. As we described in a paper that we wrote afterward (cited in the bibliography), at the operations level, there were some major gaps that seemed to trivialize the nature and character of exposure; for biological weapons, they did not provide guidance to all on the time course of exposure response relationships. Furthermore, the way in which vaccines were administered to treat the biological exposures in New Jersey were unworkable during a large-scale event, a point made by Jeanie and the major section of our paper. You cannot bring large groups of people with unknown exposures together in a distribution facility and think that you can give out medicine and prevent secondary infections at the same time. It is sort of like putting a kid who has no symptoms but is about to get the flu in a room with other children and then not expecting some of the other children to contract the flu. (Our paper is listed in the bibliography, and see our comments about PODS [Points of Drug Dispersal].) The State of NJ, Department of Health and Senior Services did review our report on TOPOFF 3 and found it useful in what they describe as "after action" evaluations.

There is still much unfinished business associated with the TOPOFF 3 exercise. When we break exposure science down to its basic elements, we know the following: if you have no contact with the toxic agent, you have no potential for disease; and if you do have contact, you must determine whether it is of sufficient intensity or of the correct duration to get sick. The trick is to understand when to be concerned and when not to be concerned.

EXPOSURE SCIENCE AND HOMELAND SECURITY

After the intense period of activity from March 1 to April 8, 2005, there was a time of relief. Alexander Lioy was born in Chicago on April 11. Since then, I have still been very active in homeland security research. However, my major interests have been the unfinished business on exposure science issues related to the attack on the WTC, the anthrax incidents, and simulated attack exercises using weapons of mass destruction or simulated accidents of highly toxic chemicals (such as chlorine), as well as exposure for deployed troops to toxic agents not linked to actual combat situations, such as a secured chemical plant.

As time has passed, I occasionally repeat the question "Have we learned anything from the WTC aerosol?" Also, is what we know now sufficient to help you get answers to these questions: How safe is it? How clean is it? When can we come back to our homes? Can people walk safely in an area? Furthermore, can we make recommendations for minimization of risk to the general population? We do not yet have the answers to many of these for acute exposure events. We must provide answers that are based on best judgment or best estimates.

It is my opinion that the country needs to engage not only the government professionals but also scientists outside the government in business and academia before, during, and after such events. We need to do a better job in preparing for exposures that can occur during the rescue and recovery periods after an event. Included are medical triage, vaccines and antidotes, and approaches to minimize additional population exposures and acute health effects. We also need better information about cleanup standards and reentry and restoration strategies. Unfortunately, for the time period during and immediately after an event, we still have many weaknesses in addressing exposure issues. I am a bit encouraged, because Robert Wood Johnson Medical School–UMDNJ, Robert Wood Johnson Hospital, and Rutgers University had vision. In late 2007, they helped a group of us form a University Center for Disaster Preparedness and Emergency Response, led by former New Jersey commissioner of health Cliff Lacy, to help address some of these and other issues. Hopefully, the fruits of our labors will start coming soon.

The issue of measuring exposure during an event is a major problem. Through 2001, our society had progressively become more inclined to obtain information, develop standards and control strategies, and establish monitoring programs for environmental situations that would help extend our lives past the age of seventy. The goal was directed to minimizing the risk for excess cancer and other long-term debilitating diseases. In addition, we were focused on investigating and minimizing low level exposure to toxic chemicals that could have developmental effects during pregnancy and the first few years of life. With our focus on these matters, though very important for environmental health and public health, it appeared to be a primary reason for our lack of being prepared to measure exposures in the immediate aftermath of an attack. As noted, we were not able to measure exposures during the attack on the WTC and provide data necessary to reduce the impact among the general population and rescue workers. We had neither the instrumentation nor the mind-set about how to focus and prioritize critical acute health issues during 9/11. However, Mike Gochfeld had a bit of a different take on the situation during exposure period 1. "I was amazed that there was no contingency for portable air sampling or alternative power for air sampling. Why could the city simply shrug its shoulders and say that the monitoring stations were down? The world is full of battery pumps" (Mike Gochfeld, pers. comm.).

Clearly, we were trying to recover lost time during the response to the attack on the WTC. In his book *Fallout*, Gonzalez indicates that the EPA was developing standards during the response, and he is critical of the agency. Maybe this was due partly to lack of communication on the way things were being done and the reason why. Maybe if there were better communication about how we have been losing ground on preparedness for acute exposure events over the prior twenty-five years, his comments would have been different. However, the following point was never made by EPA: *except for lead and PAHs, no national regulatory bodies were thinking about the development of a set of cleanup standards for dust indoors.*

As previously mentioned, a national committee for the development of AEGLs could predict what the levels of casualties would be after the release of toxic gases, but the EPA Provisional Advisory Levels, the PALs, were not on anyone's radar screen. PALs for contaminated dust still do not exist and are not even a serious consideration at the present time. The question is why? The EPA has developed them for air and water, but why not dust? How does the government expect to have adequate reentry criteria in buildings with contaminated dust or sediments (material deposited by flood waters) without them? In any case, the availability of all of these guidelines will be useful only if the monitoring instrumentation is developed to make the measurements necessary to achieve the goals of the guidelines. At the present time, there is no national research program to focus on developing these instruments. Therefore, I think that the general and effective application of these guidelines to help protect you and me are still years away.

A central question is, Are we better prepared to respond to environmental and occupational health consequences today? Yes, we are better off than on 9/11 in a number of areas, but we still have a ways to go in exposure science. The problems are not related to minor gaps; a number are related to a lack of fundamental understanding about human exposure issues. A number of items were discussed in detail in a 2007 GAO report on the second EPA Test and Clean program, which examined some of the issues that my colleagues and I have been discussing since 2002, especially in terms of respiratory protection.

However, the most important thing now is not to continue to define "who didn't do what." Let's complete unfinished business. We must start developing the instrumentation and strategies necessary to assess acute exposure or preclude additional exposures during or after cataclysmic events in the future. For example, one advance in measurement devices for asbestos has been offered to the scientific community. To quote Greg Meeker's group:

> In the years since our original study of the WTC dust, higher-spectral-resolution full-range portable spectrometers with battery operated contact light probes have been developed. These spectrometers could be used to screen thousands of spots for asbestos contamination over large, potentially-hazardous areas in a matter of days, producing data that could be spot checked with traditional evaluation methods. Automated identification software could be used to search the spectra for signs of contamination real-time, producing results immediately after each spectral measurement. (Meeker et al. 2005)

The EPA is working on remote sensors and some real-time monitors, but even today portable total or supercoarse particle dust-loading samplers are few and far between.

The USGS effort is one step forward in a process that must include a range of materials from volatile organics to the dust itself that need a more accurate set of real-time, handheld devices for situations that can identify the AEGL and the PAL levels. It is fine to have these guidelines or even standards, but without the appropriate measurement tools, they are just theoretical constructs that can be viewed for comparisons with the outputs of models, the good and the bad. Again, this is an effort that would eventually benefit from the insights of young minds from my grandsons' generation, a stimulus to complete the methods development and practical applications. A good example of a success in this area that should be reviewed by young inventors has been the development of portable radiation monitors used by police and other personnel to detect radiation releases indoors and outdoors. In contrast, real-time, handheld, or personal monitors are still needed for highly toxic organic materials and biological agents.

An important statement about the data availability after 9/11 was made by Matt Lorber in his unpublished EPA report: "These data and similar data for other contaminants underscore the importance of being able to monitor very early after such an event [the WTC attack]." It is also recognized that a major cause of the uncertainties for the evaluations presented in Lorber's report is the sparseness of information on exposures that could have occurred during that first critical week after September 11. Most of the data available were on asbestos. Regular monitoring for several contaminants of concern began mostly during the second and third week after September 11. Thus, it is clear from Lorber's analysis that the points that I have made about weaknesses in monitoring strategies and the lack of monitoring devices need to be taken to heart by all agencies. However, Phil Landrigan did provide an interesting insight about the initial measurements. He said, "They [the agencies] should have focused on more than asbestos from the beginning. But in their defense, it's like the military. They always begin by fighting the last war and then find that they have to adapt their strategy to confront new circumstances. I thought that they and you did that part well" (Phil Landrigan, pers. comm.). Let's take this point to heart, and start improving our measurement systems now.

Another note of encouragement has been the development of an air pollution monitoring plane equipped with various measuring devices. Called ASPECT (Airborne Spectral Imaging of Environmental Contamination Technology), the collaboration between the EPA and the Department of Defense led to an aircraft that can obtain chemical information from a safe distance during an emergency response situation. It provides first responders—emergency workers on the scene—with information on possible chemical releases. The problem is that there is only one, and it would not be available during the critical first hours after an event. Simple spectral monitors are needed in strategic locations "to be used quickly on the ground" around major cities and other vulnerable locations to obtain contaminant information immediately after an accident or attack.

MEASUREMENT TECHNIQUES

The United States established a network of biological monitors across the country in response to the anthrax attack. The BioWatch monitors were initially placed at locations previously selected by the EPA as air pollution–monitoring locations. This was a convenience (a simple way to get data) sample, which truly had little to do with locations that could be a high-profile location exposure caused by the release of a biological weapon. I believe that the department of Homeland Security has moved away from this strategy over the past few years to address more realistic scenarios; however, a discussion about the initial approach is still instructive for all.

To put the situation into simple terms, the EPA monitoring sites are carefully designed to look at general population impacts from emissions of typical air pollutants. For example, one typical pollutant is carbon monoxide (a chemical released from incomplete combustion), and high levels can kill people indoors when their heating system leaks in the winter time. This happens in a limited number of cases each winter, but home alarms help reduce the number of deaths. In contrast, the outdoor EPA air-monitoring sites that measure the levels of carbon monoxide in the environment do not address this issue. They are placed to monitor carbon monoxide released primarily by automobile emissions. Therefore, they measure the carbon monoxide in the center of a city or the entrance to high-traffic areas, and not near athletic fields. This approach is superb for finding the maximum amount of outdoor air pollution and consequential outdoor exposures to the general population and groups of the population susceptible to cardiac effects caused by carbon monoxide. However, do they predict the levels for the location of the population at highest risk to effects of carbon monoxide, especially the most severe effects? No. Why? The monitors do not detect carbon monoxide from leaky heating systems indoors. Carbon monoxide measurements outdoors have a specific and important purpose: detecting exposure to the pollutant when released by automobiles in a high-traffic situation. This is a reasonable and important activity in environmental health and protection, but to detect high levels indoors, you need a monitor located in your home. These are available at inexpensive prices today, and all should buy one to protect your family. However, the good news is that this multilevel exposure measurement strategy can be accomplished because we know the sources. Thus, in each case we can define approaches that measure carbon monoxide exposures and hopefully reduce the possibility of effects indoors and outdoors.

For biological weapons, the problem is more complicated because the point and time of an agent's release and the source of the release are almost impossible to know. First, in contrast to carbon monoxide, the location of the source is unpredictable. There is no routine location for a terrorist to attack. And the size of the release would also be unknown. Although the WTC was a terrorist target, it was not the target for a biological weapon in 2001. That target was the U.S. Postal Service. I know we monitor the mail now, but that is after dealing with the consequences of multiple attacks on the postal system. Would it happen there again? Think about it. A more likely location for another biological weapons attack, like anthrax, is indoors in a public place or at a large population gathering point outdoors. So, will a bioweapon monitor located at an EPA air pollution–monitoring location be adequate to respond quickly to these situations? The answer is probably no.

In summation, we need flexible chemical, biological, and radiological monitors that provide real- or near real-time measurements. Many of the WMDs are acute toxicants, so time would be truly of the essence for response, and avoidance of contact, if at all possible. These can be placed on roving vans or even better provided as roving remote sensing systems, which still need to be developed. Such systems need to be small and subject to very few physical, chemical, or biological interferences to minimize false positive readings. Both of these options, however, require significant research and validation for the different types of agents that one could expect to encounter.

Is this happening? Thankfully, in some circumstances (radiation), the answer is yes, but not to the degree necessary to handle multiple problems. The reason is we are drifting back into our comfort zone of trying to deal with long-term health effects for particulate matter and how that pollutant might affect a small number of elderly people in the general population (a point generally noted in the GAO report of 2007). Seriously, I do not understand why, because these effects pale in terms of the catastrophic consequences to the general public of being allowed back into harm's way too soon during or after a major terrorist, natural event, or accident.

It has been a recommendation of mine and others, since 2002, that we must devote the money, the time, and the effort to develop short-term monitors and rehabitation standards for events like the WTC aerosol, or a gaseous release from an accident, or a waterborne release from an accident, or any other event of consequence. These monitors must measure contaminants or indicators quickly enough so that we minimize the number of people at risk after an event. We cannot prevent the initial casualties, but we can prevent needless additional casualties, sometimes called collateral damage.

The aspect of emergency response where we still have major deficiencies and need attention in terms of strategies to define exposures and the consequences of such exposures is during and immediately after an event. Any new approaches should help identify how to minimize exposure to additional people during the period of chaos after an event. During 9/11, there was a lack of coordination and a lack of utilization of local resources and personnel. That situation did change over time. However, we did not have a strategy and still do not have a well-defined strategy to accelerate the acquisition of exposure data and information during the period of time immediately after an event. For the WTC aftermath, I defined this intense period of chaos as exposure period 1.

What we all must remember is that in contrast to a hurricane, there is little, if any opportunity to move people out of harm's way before a sudden event. I think this a point lost by many planners. Acceptance of the difficult fact that a chaotic scene will occur in the first seventy-two hours after a particular type of event will lead to more realistic understanding of what can and cannot be accomplished over this period of time. This realism about the exposures that could occur for natural versus chemical, physical, or biological releases will help local planners to determine whether local people should stay in place or evacuate after an event. I still feel that too many bureaucratic hurdles have been constructed that prevent getting the right information to the public.

There also needs to be residential (general public) and corporate training on what to do and what not to do during an event. Many corporations have developed effective contingencies, while others have not. Some corporations even have lock-down procedures. However, training of the general population is virtually nonexistent. We need you to help change this situation since planners only focus on the issues that they are forced to address first, which are preparedness of systems and responses from the response community. The general public is not involved with current "desktop" exercises. There is a commercial television program called *Homeland Security USA* that provides very good insights into many important aspects of homeland security. One day it would be great if there were an episode devoted to techniques and strategies that can be simply and effectively employed to help determine whether one should stay in place or evacuate, and the public can use or understand the tools.

During the aftermath of the collapse of the WTC, the ineffective use of the NIEHS centers as valuable resources during the rescue period was finally fixed during the recovery phase of the events. However, that was about ten days late, and it was primarily initiated by a health effects research agency. To its credit, the EPA, through the efforts of Dan Vallero, Mark Maddaloni, Alan Vetter, Gary Foley, and many others, did begin to utilize our expertise as time went on, and I think the academic communities became valuable partners in developing a flexible and integrated response. Many times the professionals outside the typical working groups, including academics and members of the general public, can help anticipate or address unusual or difficult problems. It would help if we all remember Murphy's law (if anything can go wrong, it will) to help reduce the length of the chaotic period after an event.

Over the last few years, the Department of Homeland Security has developed some Centers of Excellence, but none deal with the fundamental issues of exposure and exposure prevention or mitigation during and immediately after an event. In fact, in three visits to the Department of Homeland Security since 2005, it was only in 2008 that my colleagues and I were finally able to begin to get the message across about unfinished business. This was due to the energetic response of Dr. Charles Rodes to the previously mentioned manuscript that Pellizzari, Prezant, and I published in *Environmental Science and Technology* in 2006. He put together a fantastic session at the International Society of Exposure Science in 2007 that had some very insightful speakers, as well as an introduction from Congressman Price of North Carolina (Rodes published a summary, cited in the bibliography). Even with the unfinished business, the interest continues to wane, and only about one hundred people attended. However, the session was great, and with Charles's persistence, we finally have the ear of a member of Congress to try to move some of the issues forward and to gather the research dollars necessary to finish the business at hand. With the severe economic collapse of 2008–2009, however, national leaders' attention will likely shift from these important exposure science issues, until the next event occurs.

SOME MORE LESSONS

The one thing that the WTC has taught us is that we need to be vigilant and understand that it is not simple or easy to keep people out of harm's way. It is reasonable to have people become aware, educated, and prepared so that when they go into harm's way, they can come out alive with minimal personal injury, either immediately or at some future date, by wearing respirators or protective clothing.

We also need more training and more realistic event applications to demonstrate to people just how hard it is for an individual to go into harm's way and come out alive. As a society we have not had much training on how to deal with a large- or small-scale catastrophic event. The WTC aerosol brought us into a new era of environmental health science and the computation of the risk that can be a consequence of acute events. Sure, we are now treating the ill workers, but have we learned our lesson and provided the general public with the ability to respond personally or professionally? I think there are pockets of success in this arena, many in the private sector. But for large-scale events that affect the general public and totally disrupt communications, food supply, and/or transportation, we have yet to learn how to reduce the number of preventable exposures. We need to understand human activities and behaviors during an event that without reeducation as to the consequences deadly or harmful exposures will be high.

In July 2008, I was invited to participate in an event at the Liberty Science Center in New Jersey, a very dynamic place for kids to get enthusiastic about science and engineering. It was called "Learning from the Challenges of Our Time: 9/11, Terrorism and the Classroom." They plan to have a resource guide for educators that will include suggested curricula, sample lesson plans, information, and contact persons/institutions. I thought it was a good idea to go and learn. However, what affected me the most was the total lack of understanding or appreciation of the WTC dust exposures and scientific aspects of the aftermath of the attack. This important set of information should not be lost, and it needs a proper context for understanding how to minimize effects in the future that could be caused by major disasters. Hopefully this book will help provide a focal point for development of such a lesson plan.

I was encouraged by a couple of comments made by individuals at the meeting. They talked about developing *resiliency* among our young. It will be interesting to see how effectively that is translated into lesson plans developed about the aftermath of the terrorist attack on September 11. (The program is being pilot tested during 2009–2010, and I wish the authors well.) A sobering note was the tendency to let the WTC attack quickly drift back into the noise or put on the shelf to meet the mandated outlines for curriculums. As I mentioned in the prologue, we did that after World War II, and I think the nation forgot how truly vulnerable we are to enemies, seen or unseen.

For example, if a nuclear bomb goes off the size of Hiroshima, it would knock out quite a bit of our infrastructure, kill our communication ability, destroy transportation infrastructure, and produce exclusion zones with high radiation, which means that part of America would grind to a halt. Thus, we need education to develop *resilience* within the population, including you and me, and our families. Further training exercises are needed to remind us that if you hear something go off, do not go running out into the streets, or you may end up like some people did in Bhopal, India, during the pesticide plant explosion—dead or maimed for life. We have to be educated so that we as a society or individuals can respond effectively, rationally, in a crisis. Unfortunately, we will lose people during such an event. Some deaths and injury, such as those that occurred during the WTC attack, cannot be prevented.

The above reinforces the need for you to have a catastrophe checklist and a survival kit in your home. The main goal is to minimize what I call the additional victims or damage. From my point of view, the training has to start from the bottom up. If resiliency training is going to be effective, we have to start educating ourselves and our children in terms of how we understand the nature of materials that can hurt us badly. Thus, training associated with cataclysmic events has to be done so that we can recognize true hazards, and we can differentiate and recognize and prioritize them from other situations in a meaningful way.

We also have to understand that chemicals are all around us. Some of these chemicals are beneficial, and some toxic substances will be found at low levels in our bodies. With resiliency training and the proper understanding of harmful exposure to toxic materials, we can better understand what you and I have to do when there is a real crisis.

You can become paralyzed if there are too many apparent "crisis" situations, and then society will be unable to effectively respond to a real event. Thus, the general public has to be given information logically and in such a way that they can survive a crisis by planning, using very simple tools, and taking appropriate actions. Accomplishing such tasks requires a change in our educational systems so that children are educated to be resilient in the face of a crisis, such as a terrorist event or natural disaster, or even personal or family disasters.

Realistically, we are not going to be able minimize the risks to the point where no one is ever going to be injured or affected emotionally by a crisis or serious event. A nice goal, but this is impossible to achieve. We just cannot predict the severity of all events, even those that might be visible and reoccurring.

In one of his books, David Bates examined five environmental hazards and included mention of the 1952 London smog episode, December 5–8, that was eventually determined to have killed over eight thousand people. A passage quoted in the book is revealing: "If anyone, a few weeks ago, had suggested the possibility of a London Smog doing serious damage to cattle or other animals, submitted to its influence, he would have been looked upon as supplying himself a melancholy instance of intellectual fogginess" (quoted in Bates 1994). Thus, in a country where sulfurous smog produced by soft coal had been known for years, people never truly considered the potential of smog causing death in animals, let alone humans.

THE FIVE Rs

There are Five Rs of activities needed to deal with the exposure issues that surrounded the WTC aftermath, and each was mentioned at various points in the previous chapters. The Five Rs are *rescue, recovery, reentry, restoration*, and *rehabitation*. The first three and fifth were first presented by my colleague Dan Vallero at the meeting in 2007, and I appreciated the many discussions we had about them and his insights that were added to this section of the book. Based on my review of the overall situation at Ground Zero, I added number 4, restoration. These five issues are very important in the analysis of any catastrophic event. In fact, you can grade the response of governments and the private sector by seeing how actual exposures to hazardous conditions are addressed during each phase of an event. If we look at the attack on the WTC, the anthrax incidents, or Hurricane Katrina, where and when did the country and other entities start addressing exposure-related issues? Where did they stop? How well did they succeed? We have to go back and review what were the strengths and weaknesses of the efforts to understand and minimize exposures. Then we can evaluate the results and improve the efforts needed to minimize exposures for rescue, recovery, reentry, reconstruction, and rehabilitation. In terms of recent events, Mike Gochfeld feels that "[w]e need to ask ourselves whether we have learned lessons and are better prepared to prevent or respond to large scale disasters whether intentional, accidental, or natural. The response to Katrina suggests the answer is *no*" (Mike Gochfeld, pers. comm.). I agree with him on this point.

In the periphery around the WTC, restoration and rehabilitation have occurred. There is still, however, a lot of anxiety. We know that the Deustche Bank on Liberty Street was a major concern for a lot of people. The risk to the general population from resuspendable WTC dust in the building was low; however, the risk to recovery workers became high because of the way we were deconstructing the building. Two firefighters died in a fire that should never have occurred (WCBSTV.com, "Health Fears Loom over Deutsche Bank Building," August 25, 2007). The bank building should have been taken down years ago. There was a misplaced emphasis toward avoiding low-probability long-term exposures and health risks, and a devaluation of acute risks present at the bank site. Looking at it in another way, Mark Maddaloni of the EPA said the following: "Given the effectiveness of the WTC Residential Cleanup Program and the difficulty in demolishing 130 Liberty Street (former Deutsche Bank Building), one could only have wished that more decon and less demo prevailed" (Mark Laddaloni, pers. comm.). Thus, with a bit more emphasis on exposure prevention and education, and the use of updated versions of the EPA cleanup techniques, now over six years old, the lessons of the aftermath can be applied within restoration and rehabilitation decision-making processes in the future.

Based upon the above, and on other experiences, I believe the Five Rs are what are essential for achieving a state of equilibrium in our ability to characterize the severity of exposures that can occur and persist after an event. The concept behind the Five Rs is derived from trying to understand and answer questions related to individual and population exposure.

As a final point for your use in evaluating the response within the Five Rs, each R is attached to a period of time. Similar to the time line of exposures that occurred after the attack on the WTC, there is a progression in time for the exposures that can occur during all phases of response after an event.

For rescue, time is short; therefore, the emphasis and focus have to be directed at rescuing those who are in harm's way and trying to minimize the acute exposure that would lead to acute health outcomes among workers and the community.

The second is recovery. It requires a longer time period because what you are trying to do is recover a certain sense of normalcy in the situation and deal with the grim process of trying to retrieve bodies and trying to see where you are standing with the respect to the magnitude of the effort happening post recovery.

It's the juncture between recovery and what I call reentry that can take a longer period of time because you are now assessing the ability to utilize affected locations in a serviceable manner and do it in a way that is safe for the public. Again, this is another thing that requires education because you just can't reenter an area or reconstruct an area and hope that it's safe from toxic exposures within a very short period of time. If the WTC dust were radioactive, which it was not, buildings in downtown Manhattan compromised by the attack would not have been able to be restored and reinhabited for years (e.g., review the "exclusion zone" issues in Chernobyl, located in the Ukraine, once a region of the former Soviet Union, for some background). Thus, the longest period of time associated with an event will be associated with the restoration and rehabitation, and the exposures that one can derive from reentry or restoration. The question is, Are they minimized to the point where you can actually go back into the area and go back to work? To know this, you need reentry standards, which are still not available for most situations and compounds (e.g., residual contamination of surfaces).

Also, the issues of rehabitation really require coordination and a realistic definition of how clean is clean so that one can truly reenter to start the process that may eventually lead to rehabitation of an area. Historical comparisons with the rebuilding of Japan and Germany are always useful, as well as comparisons to recent events such as Hurricane Katrina, the earthquake in China, and the tsunami in Southeast Asia.

⑬

FINAL THOUGHTS

My story about the WTC dust and the legacy that it provides about unfinished business in the application of exposure science in emergency response has now been told. To think that settled WTC dust samples we collected in the days after September 11 would be influential in describing the progress of the aftermath, and would have such a sustained level of interest through today, is startling. However, if we take a moment to remember the images that we saw during and immediately after the disintegration of each tower, I am not surprised.

From the analyses we completed, our data on the materials in the dust, its alkaline pH, and the unusual size distribution, dominated by supercoarse particles, required time for analysis and interpretation. In the future, such time delays need to be minimized, and access to the conclusions needs to be almost real-time. The detailed characterization became important in understanding the causes of the observed acute health effects that started with the World Trade Center cough. This attempt to couple the WTC dust with health effects also led to cautions about overinterpretation of initial data. We must not forget that during the first twenty-four to seventy-two hours after the attack, there were not only high levels of particles but also high levels of unmeasured gases in the dust and smoke that were released by the fires.

The WTC dust and smoke also sparked the development of a mathematical model that reconstructed the World Trade Center plume. It has been useful in defining or estimating longer-term exposures in and around Manhattan. Maybe "World Trade Center cough" and "WTC dust" should be defined in Webster's dictionary.

Normally, this material alone would be enough to provide a good scientific story. However, little did I realize that the WTC dust would be the part of many other events in the future, which led me to write additional chapters. The observation of the lack of instrumentation and strategies to measure short-term and acute exposures became quite apparent to me and others. This point is supported by many others who noted that standards and guidelines were not available for direct comparison with the conditions encountered for acute exposure. Furthermore, early on we spent far too much time on asbestos, which is not known to be an acute toxin.

The WTC dust also showed what should be the necessary conditions of operation and expectations of a committee if it is to succeed. We should challenge those in government to review all of the transcripts and determine how to improve committees of this type in the future. A good contrast would be the success of the NASA Challenger Committee, which dealt with an equally tragic event but led to major space shuttle improvements. However, even then you still cannot totally prevent another tragedy, such as the Columbia accident that occurred in 2003.

A pleasant surprise within my analysis of the needs for exposure science in catastrophic events was the evolution of the Five Rs for exposure science. I believe these are an important contribution made by Dan Vallero, myself, and others in the group formed by Charles Rodes in 2007. The Five Rs can be used to define the exposure science needs as they relate to each specific aspect of emergency response in space and time. A fuller summary of four of the Rs is found in the report by Charles Rodes (see the bibliography). The addition of the fifth R was necessary because the path from reentry to rehabilitation can take either a short or long period of time. This period depends on the type of event, the exposures, the remaining potential for exposure, and the time and effort required to achieve rehabitation. The good news is that the Five Rs can be used as a checklist to help all of us grade the response of the public and private sector to and the level of restoration and rehabitation after a natural or man-made event or terrorist attack. So, let's put it in a form that you can use:

The Five Rs to Assess the Outcome of an Event

Rescue—the time immediately after event when the EMS and others enter an area and arc attempting to find and remove survivors.

Recovery—the time after an event when it is clear that the potential to find survivors has passed, and efforts turn toward recovering of the deceased and removal of dangerous materials.

Reentry—the time for professional assessment of the magnitude of the disaster and the development of plans for restoration.

Restoration—the time needed to rebuild or put parts of the area back into service for the community and the world. This would be dependent on how severely individual locations were affected by the event, natural or otherwise.

Rehabitation—the dates when various sectors are found to be reusable by the community and others. This could have various scales of space and time depending on how much time and effort is required to minimize exposures to the point that one can complete a restoration.

So, after an event you could take these concepts and apply them to each of the Rs (stages), and then grade the response. One could develop a checklist to help identify and understand what was done and how well the goals were managed or achieved after the event. It even could be developed as part of a lesson plan for schools.

In addition to the Five Rs is a sixth that I discussed briefly: resilience. It is very important, but it is not something that I can easily add to activities related to exposure science and other disciplines. However, without resilience, any or all of the above will at some point fail after a natural disaster, accident, or terrorist event. So in the end you have the Five Rs + One R.

I hope the Five Rs will provide you a reason to go back through the book a second time to grade how well the WTC dust exposure issues were handled, and how well other aspects of the overall event were handled by the government, businesses, and the public. Maybe doing so will give you a better understanding and appreciation about why I wrote this book for you and my grandsons.

In closing, God bless America.

AN EPILOGUE

The tenth anniversary of the attack on the World Trade Center occurs on September 11, 2011. As with other catastrophes and major historical events, the stories have been told and written (and now video archived), the wounds have begun to heal, the scars have formed, and the next generation of Americans and world citizens has been born. Further, as was the case for Pearl Harbor, Hiroshima, London, Dachau, Pompeii, and many other catastrophes noted throughout history, the images and reactions that were recorded on September 11, 2001 will remain visible to and discussed by those directly or indirectly affected until those generations pass on. We are, however fortunate that the National 9-11 Memorial Museum being constructed in NYC should provide a history lesson for generations beyond mine.

Since the publication of *Dust*, I have met many individuals who are deeply committed to ensuring that there is a sound set of history lessons that can be derived from 9-11, and been involved in some of these efforts as a result of this book. A copy of *Dust* was donated to the Library being established on the USS New York (steel beams from the Twin Towers were used in the bow of the ship). I signed and inscribed the book for the sailors. Later, I was interviewed by Mr. Cliff Chanin of the National September 11 Memorial Museum that is being constructed at Ground Zero. The interview is now part of the Museum archives, and their entire webcast series will provide information and insights on the events of 9-11, and will assess the impact of the attacks on how we view security, culture, science and medicine and politics today. Along those lines, my webcast does include further insights on my motivations for writing the book. I was also asked to donate the personal artifacts mentioned in the book (e.g. hard hat, shoes, a vial of WTC Dust etc.) to the Museum. Being involved in the museum and other commemorative activities has been a gratifying reward for the work that went into this book, which I hope will shed light on the aftermath of the catastrophe of that day.

During an interview with Ms. Alexandra Drakakis of the National September 11 Memorial Museum she described the museum and memorial and how it will use artifacts and stories to blend compassion with the history of the before, during and aftermath of the event. The National September 11 Memorial Museum will include a section for the families, a historical section on the events leading up to 9-11 and that day, and then a section on the aftermath. There will also be an archive that can be used for research and education. One area of the Memorial Museum will have what is known as the "slurry wall" which was part of the foundation of the World Trade Center. At the present time the major artifacts are being held in an 800,000 square foot Hanger, #17, and tent at Kennedy International airport. Vehicles and other items are included along with a piece of the compressed building that is composed of WTC Dust, paper, construction material etc. Ms. Drakakis is planning a visit for me to Hanger 17.

Returning to the content of *Dust*, serious questions remain; Are we recovering and are we better prepared? And have we learned any lessons related to the Five Rs? We have seen progress toward restoration (one of my Five Rs) of the WTC site; although, over the past ten years the achievements in NYC have fallen below expectations. By December, 2010, the 16-acre site is finally showing the visible signs of being rebuilt, the steel girders of two new buildings were rising above the site and good progress is being made on the reflection pools, and Memorial and Museum. The Deutsche Bank, however, continued to be removed brick by brick during the fall of 2010, a grim reminder of the WTC Dust that infiltrated many buildings on September 11, 2001. The good news is that there has been growth in the area around ground zero and many new families have re-inhabited lower Manhattan. So, progress on the fifth R is being made.

Over the year since *Dust* was published we have not learned anything new about the actual composition of the WTC Dust. I would have been very surprised by any major new findings, but there are always speculations about the nature of minor components since our analytical tools keep getting better and better at detecting small quantities of chemicals and trace materials. I still get a few requests to use the WTC Dust in toxicological studies. They have been honored, but it is always with the caution that the WTC Dust is now 10 years old and some of the more volatile components have been reduced in concentration. Other materials may have undergone changes in chemical composition even while stored at a temperature just above freezing.

I am somewhat puzzled by the fact that toxicologists are focusing on the WTC Dust in their experiments, and do not go the next step to try to simultaneously simulate the WTC Dust and fire related vapor atmospheres that was actually inhaled by all "in harm's way" during the what I described as Exposure Period 1 after the collapse. Without a simulation of both the fire related gases and the WTC Dust the results of any toxicological experiment will still provide an incomplete picture of the exposure issues that faced the rescue workers and the survivors during the first day post collapse. For the stand point of the general public, this means that we will continue to approach understanding the hazards one agent at a time or parts of a mixture, due to our inability to develop complex simulated atmospheres for toxicological studies that are a "realistic" representation of what people are exposed to in a hazardous situation.

In the future interdisciplinary research on hazards and exposures needs to be fostered to address this disconnect between laboratory studies and the realities of exposure in hazardous situations. I would call this need a goal that can bring better convergence between what is studied in the laboratory and the realities of modern environmental health problems.

Over the past two years, the legal issues of the WTC aftermath progressed to courtroom proceedings based upon on the injuries the plaintiffs said they suffered or suspected of being related to activities completed during and immediately after events of the 11th of September, 2001, primarily exposure periods 1 and 2. On June 10, 2010, Judge

Alvin K. Hellerstein, of the U.S. Federal District Court for the Southern District of New York, approved a settlement between more than 10,000 plaintiffs and the City of New York. The agreement reached to settle workers' claims against the World Trade Center Insurance Company had a monetary value of at least 625 million dollars, but the final value was dependent upon the number of plaintiffs who agreed to settle. This figure had been increased above the original settlement offer that had been rejected by the federal judge in March 2010. For the settlement to be enforced, the plaintiffs had until November 8th, extended to November 16th , 2010 to have 95 percent of the plaintiffs approve. 95.1 percent plaintiffs agreed to settle. In the end, it remains to be seen whether or not the settlement yields some degree of closure to those who sued for WTC related injuries.

More generally, going shopping did not bring the war on terrorism to conclusion, but at least the Yankees won another World Series in 2009.

Of concern to many professionals are the continuing deficiencies in and doubts about our anti-terrorism program. After the near miss from the attempted Times Square bombing during the spring of 2010, coming about 5 months after the attempted Christmas day attack on a commercial aircraft, a poignant interview with former New Jersey Governor Tom Kean published in the *Star Ledger* May 30, 2010. His final points in the article were that "[t]he FBI is still not fixed, and Homeland Security spends way too much of its time reporting to the 100 congressional subcommittees that claim some type of jurisdiction," Kean claimed. "We can't count on the terrorists being incompetent forever." Thus, part of the answer to my question is, no. This was further emphasized by the attempt to bomb cargo planes bound toward Chicago this past fall. The terrorists continue to probe for points of weakness.

In a discussion by Commander, U.S. Special Operations Command Adm. Eric T. Olson, after his March 16, 2010 meeting with the Senate Armed Services committee, he indicated that the big threat is extremists acquiring chemical, biological or nuclear arms. (Tony Capaccio, Bloomberg News, Star Ledger, page7, August 29, 2010) Thus, the need to understand and employ the Five Rs as they relate to human exposures in planning is still critical.

I was pleasantly surprised by an article authored by J. M. McNamara and M. Rastakia, "Exposure issues in responders to disasters—offering some ideas and lessons learned from the response to the World Trade Center 911 tragedy."[1] They discussed *Dust*'s Five Rs as a new paradigm for responders, and how it can be used as "an interval approach" to dealing with an aftermath. Thus, the messages in the book on rescue, reentry, recovery, restoration and rehabilitation are beginning to have some practical impact in discussions about Emergency Response, I hope these concepts translate to action. My message is also being reinforced by an article I published in *Military Medicine* in 2011 on how the lessons we learning from 9-11 can be used to help protect our troops in non-combat situations. This article was a result of a workshop discussion on the content of *Dust* and lessons learned from 9-11 at a symposium on Assessing Potentially Hazardous Environmental Exposures among Military Populations. It was held at the Uniformed Services University of Health Sciences, Bethesda, Maryland, in May 2010.

Since there will always be consequential accidents and other tragedies, it is worth recounting some of lessons from the WTC aftermath. Both volunteers and professionals need to be trained to handle and understand the requests for personal protective equipment, need to wear and use such equipment, and need to follow established procedures. Based upon the experiences during the WTC aftermath, the Five Rs of an accident or attack require that the use of personal protective equipment (PPE) or procedures for conducting specific activities must be constantly reinforced by the situation managers and supervisory personnel. Since the attack, Dan Vallero and others at EPA have recommended a manifest system for the future.[2] Finally, individuals must continue to exercise common sense about when to use or not use such equipment. Clearly, during 9-11, the general announcements by officials and information provided by the media did not work or were misinterpreted.

THE 5 RS AND GULF OIL SITUATION DURING THE SUMMER OF 2010

This past summer we had a new catastrophe, the Gulf Oil fire and leak. I did get involved with the aftermath, though not to the degree post-9-11. It did, however, provide the opportunity to examine how the response related to the Five Rs in Chapter 12, and the use of personal protection.

Personal Protection

As I've described in these pages, using personal protective equipment is extremely important for those volunteering during any given situation. Obviously, during the initial phases of rescue, personal protection equipment distribution and training will not be possible, but as one moves from initial rescue operation and then into the phases of re-entry and recovery this should be a priority. The Deepwater Horizon Gulf Coast Oil leak of 2010 provided a rare situation where volunteers could be trained and outfitted properly to address the issues of intervention along the coast while removing new or aging leaked oil. This was feasible because the leak was not a one-day release; it lasted for over 85 days. During a June 22nd presentation at an Institute of Medicine meeting on the Gulf in New Orleans, LA I made this point multiple times. The talk and other presentations have been video archived by the Institute of Medicine[3]. The good news was that the National Occupation Health and Safety Administration, under Dr. John Howard, and OSHA established a set of criteria for jobs and activities associated with the spill and clean up, and a list of protective equipment and the requirements.[4] These efforts appeared to have been effective in the Gulf region, which should reduce the number of injuries and health effects that will be experienced by the workers and volunteer or National Guard cleanup workers. Further, the framework established by NIOSH/OSHA should be translatable to other situations.

Aged Oil—Skimmers and Recovery

I also reviewed where the country was in its response to the spill based upon the Five Rs, outlined in chapters 12 and 13, again noting my collaborator Dan Vallero, EPA. My message was that the rescue phase was complete; tragically, eleven workers died during explosion and fire. Re-entry into the Deepwater Horizon site was successful, and operations on the leak started in April, 2010. Unfortunately, as of my June talk the leak continued. By July 27–29, 2010, days my EOHSI Division was sampling in the Gulf, the leak had a temporary cap. A permanent solution was finally achieved by summer's end.

Initial attempts at Recovery for the Gulf shore communities and businesses were difficult. During the first sixty days after the fire there was not much recovery. In fact, efforts to prevent oil and dispersant (CO-REXIT) from reaching shore were a failure, crippling recovery efforts. This had been graphically displayed using interactive Graphic Information Systems (GIS) by many. In contrast to the WTC, the leak continued every day, twenty-four hours a day for over eighty days. Thus, the recovery problems were renewed every day. Further, the potential for environmental and human health effects expanded as the aging oil/dispersant mixture moved on shore and/or inland. The difference each day was weather, and its effect when the oil moved on shore. The government and BP should have understood this more quickly and been aggressive in recovering oil. If we had used all available resources early on, skimmers, barges, and other devices to pick up the oil, the conditions on the beach and in the interior could have been better. Not perfect, but better.

There was a lack of media attention through June 2010 (first sixty days) on why the Administration and BP were slow in employing skimmers etc to improve recovery efforts. The obvious question was: Why were the skimmers and barges delayed or prevented from going into the Gulf? Of concern to EPA was that some the superior European skimming technology could not meet U.S. water quality standards. Dutch vessels, for example, continuously suck up vast quantities of oily water, extract most of the oil and then spit overboard vast quantities of nearly oil-free water. "Nearly oil-free isn't good enough for the U.S. EPA, who have a standard of 15 parts per million—if water isn't at least 99.9985 percent pure, it may not be returned to the Gulf of Mexico."[5] I suggest that EPA would have found that the un-skimmed aged oil/dispersant in the Gulf near shore caused much more significant violations of water quality standards.

On the floor of U.S. Senate, Sen. George LeMieux, R-Fla., pointed to a Coast Guard map detailing more than 850 skimmers available in the southeastern United States—and more than 1,600 available in the continental United States. The Oil Pollution Act of 1990 requires regions to have minimum levels of equipment such as boom and skimmers, making it difficult for every oil-fighting resource to be directed to the Gulf of Mexico.[6] Under the National Response Plan (NRP), the Federal On-Scene Officer (OSC) has the authority to use many resources to deal with a significant event. The National Response Plan had been significantly upgraded because of the events of 9-11. It is a pity that it was not implemented effectively to reduce impact on the Gulf shores. Rear Adm. Mary Landry, and then Coast Guard Rear Adm. James Watson after May, 2010 were the OSCs.

According to the NRP, section §**300.322** authorizes the OSC to determine whether a release poses a substantial threat to the public health or welfare of the United States based on several factors, including the size and character of the discharge and its proximity to human populations and sensitive environments. In such cases, the OSC is authorized to direct all federal, state, or private response and recovery actions, and can obtain support from other federal agencies or special teams. In the absence of or need to improve the actions of a responsible party (BP), the government established the methods and guidelines that are used to eliminate the threat during major spills or in responding to significant spills or spills of national significance. As operations are completed, if there are tasks that need to be done to protect the environmental and public health, the OSC can use the "most effective means possible" to safely address the problem, in this case a leak. Why didn't the OSC seize control and require use of all available skimming resources early during the event? Rear Admiral Landry still owes the nation answers.

A Deepwater Horizon Incident Joint Information Center update on July 17th (titled "'A Whale' operational review") stated that "since early June, at the direction of National Incident Commander Admiral Thad Allen, the number of skimmers fighting oil in the Gulf has been increased more than fivefold to 593 as of today. There are currently more than total 6,800 vessels responding on site, including skimmers, tugs, barges, and recovery vessels to assist in containment and cleanup efforts—in addition to dozens of aircraft, remotely operated vehicles, and multiple mobile offshore drilling units." I leave to you to decide whether or not this response time as reasonable for the residents along the Gulf of Mexico. The following statement from Plaquemines Parish President Billy Nungesser defined what was needed for recovery: "We want all the skimming vessels in the world deployed."

Finally, after 9-11 US EPA received recommendations based upon my work, see chapters 2 and 3, and others about the need to have the technology available to measure "super coarse" particles in addition to fine particles in specific types of situations. This was important monitoring tool needed the Gulf. However, EPA only measured fine and coarse particles, which are not high priority when faced with oil and dispersant being released by the wave action on a beach. Much larger particles would lead to the exposures of concern. A portable "super coarse" monitor is still needed for the EPA monitoring tool box. One local beach supervisor said that it is hard to comprehend the lack of understanding that oil contaminated sea spray is the highest sampling priority after volatile organics. Unfortunately, the latter were measured only sporadically in the Grande Isle area. The Nation's scientists and engineers still requires a serious review of the types and number of monitoring and sampling tools that are available for disasters, and how to select those that can and cannot provide valuable information in a specific situation to prevent further exposures or address a number of possible events. Thus, the efforts by the panel put together by Dr. Rodes, see chapter 12, still need to be effectively implemented through legislation in congress and then integration into various agencies.

EOHSI work in the Gulf

We took measurements in the Gulf, and it was an education. We saw the crews properly outfitted with personal protection cleaning the sand on the beaches where aged oil had deposited after traveling from the leak site. On Elmer's Island, Louisiana we collected samples of recently deposited oil, aged tar balls, and sand that were contaminated with oil. At the time, the beaches were still closed on Grande Isle and Elmer's Isle. I must thank Ralph Portier of Louisiana State University (brother of Chris Portier from 9-11) for getting us to the Island. Just like 9-11, the bureaucracy and security had become almost insurmountable. Two weeks earlier our technician, Kris Mohan, also had been in Pensacola, Fl. and had a much easier time collecting samples through the assistance of the MOTE Marine Laboratory in Sarasota, Florida.

The characteristics of the fresh oil and aged oil samples, shoreline water samples, and surface wipes and air samples were different. We will be reporting the final results, and conclusions over time. One initial observation was that each type of beach, marsh etc. will have to be evaluated for short and long term recovery issues because of the slowness in preventing the oil from reaching shore. I hope we learn the lessons that can prevent exposures during recovery.

The latest disaster is the earthquake and tsunami in Japan, and subsequent overheating nuclear fuel rods and release of radioactivity at the six-reactor Fukushima plant in northeastern Japan, 150 miles north of Tokyo. The three events provide another opportunity to evaluate the effectiveness of all stakeholders in completing the Five Rs. In this case there will be both short term (e.g. acute health outcomes and safety) and long term (e.g. entombing reactors, and rehabitation) issues, and the ability to restore different sectors will be evaluated for success or failure over many years.

A thank you goes out to all I have met since the publication of *Dust*. I continue to learn more because of what each of you said during our conversations.

Paul J. Lioy

NOTES

1. *Tak/Response*, San Jose, Ca, September 14–16, 2010. www.takresponse .com/index/disaster-news/exposure issues_in.html.

2. www.isesweb.org/Meetings/Docs/ISEA2007_wtc_lessons_learned_ vallero_addon.pdf

3. www.iom.edu/oilspillhealth

4. Available at http://www.osha.gov/oilspills/oil_ppematrix.html and http:// www.cdc.gov/niosh/topics/oilspillresponse/pperecsumm.html, respectively

5. *Avoidable Catastrophe*, Lawrence Solomon, Financial Post (CA), June 26, 2010

6. Red tape keeps prized oil-fighting skimmers from Gulf, coastline, Chris Kirkham, the Times-Picayune, June 28, 2010

APPENDIX A:
PRESENTATION OF THE DETAILED DATA SET ON THE COMPOUNDS MEASURED IN THE WTC DUST SAMPLES

Table A.1. General Characteristics of Settled Dust and Smoke Samples (Percent by Mass) from the First Two Days after the Collapse and Fires of the WTC

Sample		Street	
	Cortlandt	Cherry	Market
Color	Pinkish Gray	Pinkish Gray	Pinkish Gray
pH	11.5	9.2	9.3
Nonfiber (cement/carbon; %)[a]	50.0	49.2	37.0
Glass fiber (%)[a]	40.0	40.0	40.0
Cellulose (%)[a]	9.2	10.0	20.0
Chrysotile (%)[a]	0.8	0.8	3.0
Aerodynamically separated sample (% mass)			
<2.5 um diameter	1.12	0.88	1.30
2.3–10 um diameter	0.35	0.30	0.40
10–53 um diameter	37.03	46.61	34.69
>53 um diameter	61.50	52.21	63.60
Sieved sample (% mass)			
<75 um diameter	38.00	30.00	37.00
75–300 um diameter	46.00	49.00	42.00
>300	16.00	23.00	21.00
Anions/cations (μg/g)[b]			
Fluoride	220	70	ND
Chloride	800	270	220
Nitrate	330	ND	ND
Sulfate	41,400	35,200	42,100
Calcium	18,200	14,000	17,700
Sodium	400	200	130
Potassium	60	170	270

Source: Lioy et al., *Environmental Health Perspectives* 110, no. 7 (2002): 703–714. Used with permission.
Note: ND = not detectable
[a] Values reported to L. C. Chen by the Ambient Group, TNC (New York), measured by polarized light micros-copy (400–450×).
[b] The units reported here are correct. The units reported in the original article were ng/g, which is incorrect.

Table A.2. **Concentrations of Elements (in ng/g, Dry Weight) in the Three Settled Dust and Smoke Samples**

| Element | Street | | |
	Cortlandt	Cherry	Market
Li	26,800	22,650	29,520
Be	3,292	2,638	3,754
Mg	110,300	118,300	179,000
Al	814,700	558,800	908,700
Ti	1,717,000	1,485,000	1,797,000
V	40,370	33,890	42,610
Cr	182,000	142,600	171,500
Mn	781,400	565,100	828,100
Co	8,316	7,230	10,460
Ni	41,140	42,040	47,290
Cu	133,500	336,300	325,600
Zn	1,718,000	2,555,000	2,992,000
Ga	30,300	26,990	31,010
As	2,464	2,792	2,613
Rb	21,390	18,630	21,710
Sr	691,000	478,900	720,800
Ag	2,565	1,945	2,247
Cd	5,695	8,454	7,459
Cs	1,165	1,085	1,327
Ba	365,300	370,000	406,500
Hg	ND	ND	ND
Tl	905	1,954	1,290
Pb	142,400	483,500	289,200
Bi	1,087	1,405	1,466
U	4,117	3,920	4,213

Source: Lioy et al., *Environmental Health Perspectives* 110 (2002): 703–714. Used with permission.

Table A.3. Concentrations of Pesticides, PCBs, and Selected PAHs (ng/g) in the Three Settled Dust and Smoke Samples

Compound	Street		
	Cortlandt	Cherry	Market
Pesticides			
Hexachlorobenzene	1.9	0.9	1.2
Heptachlor	ND	ND	ND
4.4 DDE	1.3	2.1	3.0
2.4 DDT	ND	ND	ND
4,4 DDT	ND	ND	ND
Mirex	ND	0.8	ND
Total Chlordanes	3.1	5.6	3.7
PCBs[a]			
Total PCBs (without 8 and 5)	631	562	723
Total PCBs (with 8 and 5)	659	589	753
Selected PAHs			
Fluorene	6,800	2,620	32,200
Phenanthrene	44,100	22,300	32,100
Fluoranthrene	40,300	13,700	32,600
Benxol[a]pyrene	23,000	12,100	19,300
Benzol[b+k]fluoranthane	36,600	15,600	29,500
Total PAHs (40 compounds)[a]	383,300	218,100	376,100

Source: Lioy et al., *Environmental Health Perspectives* 110, no. 7 (2002): 703–714. Used with permission.
Note: ND = not detected
[a]PCB concentrations are the sum of 68 congeners (66 congeners in the case of sum PCB without IUPAC congeners 5 and 8).

Table A.4   Concentrations of Semivolatile Hydrocarbons (µg/g) Found in the Three Dust and Smoke Samples (Includes Only Compounds, 5 µg/g in Concentrations)

Compound name	Street		
	Market	Cherry	Cortlandt
2-Hexyl-1-decanol	ND	37.2	37.4
1-Dodecanol,2-methyl-, (S)-	ND	ND	6.8
1H-1,2,4-Triazole, 1-ethyl	ND	12.1	ND
1H-Indene, 1-(phenylmethylene)-	ND	ND	10.0
1H-Pyrrole-3-propanoic acid, 2,5-dihydro-4-methyl-2,5-dioxo	9.0	ND	ND
1-Hexadecanol, 2-methyl	ND	19.7	ND
1-Hexyl-2-nitrocyclohexane	27.0	ND	ND
1-Hydroxypyrene	14.1	ND	ND
2,3-Dimethyl-1-pentanol,	20.4	ND	ND
1-Pentacontanol	ND	ND	27.7
4-Methyl-2-propyl-1-pentanol	ND	ND	27.5
1,2,3-Triphenyl-3-vinyl-cyclopropene	ND	ND	24.5
2-Benzylquinoline	ND	ND	18.8
2-Methylnaphthalene	ND	ND	5.1
2,3-Dihydrofluoranthene	13.4	ND	ND
2-(3'-Hydroxyphenylamino) -5-methyl-4-oxo-3,4-dihydrophyrimidine	ND	ND	41.4
3-Methoxycarbonyl-2-methyl-5-(2,3,5-tri- O-acetyl-beta-d-ribofuranosyl)	32.1	ND	ND
3,3'-Dochlorobenzidine	10.0	ND	ND
3,4-Dihydrocyclopental(cd)pyrene (acepyrrene)	35.5	ND	ND
4-Hydrozymandelic acid-TRITMS	26.8	ND	ND
7-Methyl 3,4,5(2H) tetrahydroazepine	ND	ND	10.5
Benzo(c)fluorine	39.4	ND	ND
9H-Fluorene, 9-(phenylmethylene)	13.8	ND	ND
9,10 Anthraquinone	21.4	ND	11.5
11H-Benzo(a)fluorene	19.6	ND	ND
11H-Benzo(b)fluorene	33.3	ND	ND
12-Acetoxydaphnetoxin	ND	ND	8.0
1-Methylanthracene	8.9	ND	ND
Auraptenol	ND	ND	13.5
Benz[a]acridine, 10-methyl-	ND	ND	7.2
Benz[a]acridine	ND	12.4	ND
Benzamide, N-acetyl-	ND	ND	22.9
Benzene, 1,1'-(1,3-butadiyne-1,4-diyl)bis-	79.8	ND	ND
Benzimidazo[2,1-a] isoquinoline	ND	17.3	ND
Benzo[a]acridine	9.3	ND	ND
Benz[a]anthracene	61.0	ND	ND
Benzo[b]fluoranthene (benz[e]acephenanthylene)	49.9	ND	ND
Benzo[b]naphthol[2,3-d]furan	ND	ND	28.2
Benzo[c]phenanthrene	ND	ND	43.7
Benzo[h]quinoline	ND	ND	5.9
Benzyl butyl phthalate	ND	ND	94.1

Table A.4   (Contd.)

Compound name	Street		
	Market	Cherry	Cortlandt
Biphenyl	6.5	ND	ND
Carbazole	28.6	8.1	22.1
Dimethylcyanamide	ND	ND	14.4
Cyclohexanemethanol	ND	ND	16.8
Cycloate	ND	32.4	46.0
Diisobutyl phthalate	27.5	ND	ND
Di-*n*-butyl phthalate	12.6	ND	ND
Dibenzofuran	14.5	ND	9.2
Dibenzothiophene	ND	ND	12.7
Dibutyl phthalate	16.5	14.6	19.7
Dicyclohexyl phthalate	ND	77.8	ND
Didodecyl phthalate	80.0	ND	ND
Diethyl phthalate	31.7	ND	ND
Dihydrogeraniol	51.4	ND	ND
Droserone (2,8-dihroxy-3-methyl-1,4-naphthoquinone)	9.8	ND	ND
(E)-2-(6-nonexnoxy)-tetrahydropyran	14.5	ND	ND
Ether, hexyl pentyl	31.6	31.3	ND
2,4-Dimethylheptane	ND	28.5	ND
2,3,4-Trimethylhexane	12.2	15.6	12.7
2,4-Dimethylhexane	ND	ND	13.3
3,3-Dimethylhexane	ND	14.1	30.8
Hexyl *N*-butyrate	ND	ND	8.0
Methyl alpha-ketopalmitate	ND	ND	37.3
Metribuzin	22.1	ND	ND
Monobutyl phthalate	ND	ND	62.4
n-Octane	43.3	ND	ND
Naphthalene, 1-(methylthio)-	ND	ND	7.5
Naphthalene, 1,3-dimethylene	5.3	ND	ND
Nefopam	10.2	ND	ND
Octane	ND	6.9	ND
Pentanioc acid,4,4-dimethyl-3methylene-,ethyl ester	ND	19.5	ND
1-Methylphenanthrene	ND	ND	10.5
4-Methylphenanthrene	ND	ND	12.9
Phthalate	ND	6.9	ND
Phthalic acid, 2-hexyl ester	ND	47.9	ND
Prometryn (caparol)	10.7	ND	ND
4,4'-Biphenyldicarbonitrile	ND	ND	19.6
Chrysene	ND	18.2	ND
1-Azabicyclo[2.2.2]octan-3-one	ND	12.3	ND
Venolate (vernam)	ND	ND	14.9
Xanthene	9.5	ND	ND

Source: Lioy et al., *Environmental Health Perspectives* 110, no. 7 (2002): 703–714. Used with permission.
Note: NA, not available; ND, not detected.

Table A.5. Concentrations (ng/kg) of PCDDs and PCDFs in the Three Settled Dust Samples

Compound	Street		
	Cortlandt	Market	Cherry
PCDDs			
2,3,7,8-TCDD	7.00	5.81	6.53
1,2,3,7,8-PentaCDD	29.4	6.01	7.05
1,2,3,4,7,8-HexaCDD	32.2	4.93	4.95
1,2,3,6,7,8-HexaCDD	35.0	16.6	18.0
1,2,3,7,8,9-HexaCDD	38.5	19.0	18.8
1,2,3,4,6,7,8-HeptaCDD	158	304	325
1,2,3,4,6,7,8,9-OctaCDD	1,450	3,630	3,410
PCDFs			
2,3,7,8-TetraCDF	78.2	194	221
1,2,3,7,8-PentaCDF	40.3	39.4	42.7
2,3,4,7,8-PentaCDF	54.5	77.3	85.0
1,2,3,4,7,8-HexaCDF	57.4	46.0	46.4
1,2,3,6,7,8-HexaCDF	46.1	46.0	48.0
1,2,3,7,8,9-HexaCDF	28.7	24.4	4.95
2,3,4,6,7,8-HexaCDF	39.5	37.0	37.9
1,2,3,4,6,7,8-HeptaCDF	91.6	171	177
1,2,3,4,7,8,9-HeptaCDF	40.1	21	31.1
1,2,3,4,6,7,8,9-OctaCDF	118	171	182
2,3,7,8-Dioxin total equivalents	104	96.1	103

Source: Lioy et al., *Environmental Health Perspectives* 110, no. 7 (2002): 703–714. Used with permission.

Table A.6. Concentrations of Some Brominated Flame Retardants in the Dust and Smoke Released by the Collapse of the WTC (μg/kg, Dry Weight Basis)

Street	BDE47	BDE100	BDE99	BDE153	BDE154 + PBB153	BDE209
Cherry	146	74.1	234	45.9	219	1,330
Cortlandt	107	64.2	155	42.0	305	2,660
Market	174	51.1	293	53.4	243	2,330

Source: Lioy et al., *Environmental Health Perspectives*, 110, no. 7 (2002): 703–714. Used with permission.
Note: BDE47, 2,2',4,4'-tetrabromodiphenyl ether; BDE99, 2,2',4,4',5-pentabromodiphenyl ether; BDE100, 2,2',4,4',6-pentabromodiphenyl ether; BDE153, 2,2',4,4',5,5'-hexabromodiphenyl ether; BDE154, 2,2',4,4' ,5,6'-hexabromodiphenyl ether; BDE209, 2,2',3,3',4,4',5,5',6,6'-hexabromodiphenyl ether; PBB153, 2,2',4,4' ,5,5'-hexabromobiphenyl.

Table A.7. General Characteristics of Settled Dust and Smoke Samples Indoors for Buildings near Ground Zero

Address	Liberty Street											Trinity Place		
Floor	8		5		4					2		7		10
Location	LR & BR	Library	Hall	Baseboard	Floor	Hall & LR	High Chair	BR	Front Room	Front Room	Entryway	Walkway	Office	Office Window
Color	Gray/tan	N/A	gray/tan	N/A	gray/tan	gray/tan	tan	N/A	tan	N/A	gray/tan	gray/tan	gray/tan	N/A
pH	11	N/A	11	N/A	11	11	11	N/A	11	N/A	11	8	10	N/A
Chrysotile fibers (% volume)	<1	N/A	<1	N/A	<1	<1	<1	N/A	<1	N/A	<1	<1	<1	N/A
Sieved sample (% mass)														
>1000μm	1	N/A	1	N/A	2	1	3	N/A	6	N/A	1	28	5	N/A
53–1000 μm	72	N/A	73	N/A	69	67	74	N/A	72	N/A	69	59	67	N/A
<53 μm	27	N/A	27	N/A	28	33	23	N/A	22	N/A	31	13	28	N/A
Aerodynamically separated sample (% mass)														
<2.5 μm	0.6	0.4	0.7	0.4	0.8	0.8	0.6	0.6	0.5	0.5	0.6	0.2	0.7	0.6
2.5–10μm	1.3	1.2	2.3	1.4	1.5	2	1.8	2	1.5	1.8	2.2	0.5	1.6	1.5
10–53 μm	78.5	51.4	64.7	60.6	63	72.1	59	61.4	50.9	42.4	71.6	20.1	58.2	56.3
>53 μm	19.6	47	32.3	37.6	34.7	25.2	38.6	36	47.1	55.3	25.3	79.1	39.5	41.3

Source: Adapted from Lioy presentations and data files.
Note: LR = living room; BR = bedroom; N/A = not available; ND = not detected.

Table A.8. Concentrations (ng/g) of Elements Found in the Indoor Dust and Smoke Samples

Address	Liberty Street			Trinity Place			
Floor	8	5	4	2	7	8	10
Li	16,000	15,800	13,000	ND	12,200	10,200	15,900
Be	1,900	1,690	2,140	1,490	ND	ND	1,500
Ti	1,120,000	1,000,000	850,000	945,000	971,000	834,000	1,010,000
V	20,300	21,100	16,700	16,200	13,800	24,400	18,500
Cr	79,700	82,300	60,300	55,000	60,100	51,900	66,500
Mn	654,000	573,000	446,000	499,000	570,000	415,000	565,000
Co	3,780	4,560	3,680	3,560	3,040	4,120	3,350
Cu	172,000	162,000	99,100	94,000	348,000	236,000	139,000
Ga	9,470	9,710	7,770	7,890	6,780	6,960	8,850
As	2,330	3,380	3,290	3,490	2,600	3,790	2,450
Rb	10,500	10,400	8,380	9,310	9,480	7,960	9,640
Sr	562,000	490,000	368,000	416,000	433,000	442,000	479,000
Ag	1,600	1,080	1,550	956	153	1,570	781
Cd	4,060	3,230	3,120	5,300	3,130	5,340	5,400
Cs	ND	ND	ND	ND	ND	ND	ND
Ba	222,000	212,000	170,000	178,000	149,000	166,000	219,000
Hg	ND	ND	ND	ND	ND	ND	ND
Tl	ND	ND	ND	ND	ND	ND	ND
Pb	158,000	174,000	112,000	126,000	51,600	246,000	187,000
Bi	ND	ND	ND	ND	ND	ND	ND
U	ND	ND	ND	ND	ND	ND	ND

Source: Adapted from Lioy presentations and published manuscripts.
Note: ND = not detected. Average is shown for multiple-sample IDs.

Table A.9 Concentrations (µg/g) of the Most Frequently Found Semivolatile Hydrocarbons in the Indoor Dust and Smoke Samples

	Address										
		Liberty Street						Trinity Place			
Compound Name	Floor	8	5	5	4	4	2	2	2	7	10	10
1-Decanal, 2-hexyl-		ND	3.07	3.99	ND	1.9	3.09	ND	ND	ND	ND	ND
Acenaphthene		ND	3.4	ND	0.1-5	5.23	4.06	ND	0.1-5	17	ND	ND
Anthracene		ND	ND	9.17	0.1-5	20.4	16.2	2.25	1.76	13.7	1.82	0.09
Benz(c)acridine		ND	ND	ND	ND	ND	ND	0.41	0.38	4.72	0.05	ND
Benz(a)acridine		ND	4.97	3.28	ND	ND	11.80	ND	14.10	ND	ND	ND
Benzo(a)pyrene		ND	5.04	5.48	ND	7.68	3.62	ND	13.5	ND	1.88	0.05
Benzoic acid,p-tert butyl		2.68	2.73	20.2	0.1-5	2.50	1.80	ND	0.1-5	86.9	ND	ND
Benzyl butyl phthalate		ND	9.43	9.89	ND	8.94	ND	1.53	ND	30.3	ND	ND
Carbazole		ND	0.77	0.94	ND	2.02	0.77	ND	ND	ND	ND	ND
Dibenzofuran		ND	1.97	2.13	0.1-5	2.81	2.44	ND	0.1-5	20.3	ND	ND
Dibutyl phthalate		1.75	0.29	22.9	ND	9.52	1.47	0.85	ND	8.32	3.99	0.10
Diethyl phthalate		ND	ND	2.85	ND	3.43	ND	ND	0.1-5	ND	1.61	0.03
Fluoranthene		ND	15.70	9.15	ND	30.9	11.00	ND	ND	23.1	ND	0.18
Fluorene		ND	2.89	2.66	ND	4.52	2.33	0.5-1	0.1-5	18.2	ND	ND
Phenanthrene		ND	ND	9.17	0.1-5	2.99	2.50	ND	2.62	7.02	1.82	ND
Pyrene		ND	ND	14.2	ND	22.5	ND	1.66	11.6	11.3	ND	ND
Thebenidine		ND	21.0	4.56	0.1-5	ND	23.9	0.1-5	25.9	ND	2.13	0.08

Source: Adapted from Lioy et al. data files.
Note: ND = not detected.

Table A.10. Concentrations (ng/g) of Dioxins/Furans Found in the Indoor Dust and Smoke Samples

Address	Liberty Street			Trinity Place	
Floor	5	4	2	7	10
Dioxin (ng/kg)					
2,3,7,8-tetrachlorodibenzo-p-dioxin	9.6	8.9	7.4	3	20
1,2,3,7,8-Pentachlorodibenzo-p-dioxin	37.2	24.9	19.1	5	30.6
1,2,3,4,7,8-Hexachlorodibenzo-p-dioxin	37.2	24.9	16.7	5	15
1,2,3,6,7,8-Hexachlorodibenzo-p-dioxin	37.2	24.9	23.2	7	35.7
1,2,3,7,8,9-Hexachlorodibenzo-p-dioxin	37.2	24.9	28.2	9.1	55.7
1,2,3,6,7,8-Heptachlorodibenzo-p-dioxin	334	385	313	242	500
1,2,3,4,6,7,8,9 Octachlorodibenzo p dioxin	3755	5375	3630	2620	3765
Furan (ng/kg)					
2,3,7,8-Tetrachlorodibenzo-p-furan	261	331	278	123	1119
1,2,3,7,8-Pentachlorodibenzo-p-furan	50.1	46	38.3	14.9	117
2,3,4,7,8-Pentachlorodibenzo-p-furan	80.2	72.3	56.5	26.4	158
1,2,3,4,7,8-Hexachlorodibenzo-p-furan	50.8	50.4	53.1	20.4	89.4
1,2,3,6,7,8-Hexachlorodibenzo-p-furan	49.4	45.8	37.9	15.9	25
1,2,3,7,8,9-Hexachlorodibenzo-p-furan	37.2	24.9	15.1	5	8.2
2,3,4,6,7,8-Hexachlorodibenzo-p-furan	47.5	38.5	37.1	13.8	84.4
1,2,3,4,6,7,8-Heptachlorodibenzo-p-furan	199	204	157	68.8	342
1,2,3,4,7,8,9-Heptachlorodibenzo-p-furan	41.5	29.9	27.6	5	54.9
1,2,3,4,6,7,8,9-Octachlorodibenzo-p-furan	223.	275	184	78.1	278
2,3,7,8-Dioxin total equivalents	96.5	100	87.3	39.2	293

Source: Adapted from data files of Lioy et al.
Note: Average is shown for multiple IDs.

APPENDIX B:
ORIGINAL MEMBERS OF
THE WTC TECHNICAL PANEL

Paul Gilman—Office of Research and Development, U.S. Environmental Protection Agency

Paul J. Lioy, Ph.D. **(Vice Chair)**—UMDNJ, Robert Wood Johnson Medical School

Patricia Clark—Occupational Health and Safety Administration

Commander Peter W. Gautier—Gulf Strike Team, U.S. Coast Guard

Jessica Leighton, Ph.D., M.P.H.—New York City Department of Health and Mental Hygiene Representative

Morton Lippmann, Ph.D.—New York University School of Medicine

Steven Markowitz, M.D.—Queens College, City University of New York

Gregory P. Meeker—Electron Microbeam Laboratory, U.S. Geological Survey

David M. Newman, M.A., M.S.—New York Committee for Occupational Safety and Health

Frederica P. Perera, Ph.D.—Columbia University Mailman School of Public Health and Columbia Center for

Children's Environmental Health

David J. Prezant, M.D.—Albert Einstein College of Medicine and New York City Fire Department

Krish Radhakrishnan—New York City Department of Environmental Protection

Sven E. Rodenbeck, Sc.D.P.E., DEE—Division of Health Assessment and Consultation, Agency for Toxic Substances and Disease Registry

Jeanne Mager Stellman, Ph.D.—Mailman School of Public Health, Columbia University

Catherine McVay Hughes (ex officio)—Community Liaison

Joseph Picciano (ex officio)—Federal Emergency Management Agency, New York City office

Claudia Thompson Ph.D. (ex officio)—Center for Risk and Integrated Sciences, Division of Extramural Research and Training, National Institute of Environmental Health Sciences

Marc Wilkenfeld, M.D. (ex officio)—board-certified occupational/environmental physician in New York City

APPENDIX C: COMMENTS ON WTC SIGNATURE STUDY AND PEER REVIEW FROM GREG MEEKER, PAUL LIOY, AND MORT LIPPMANN, NOVEMBER 3, 2005

The peer review of the Final Report on the World Trade Center (WTC) Dust Screening Method Study appears to be a thorough review of the materials provided to the peer reviewers by EPA. There are some excellent comments that should be considered before any final decision is made on whether or not to use a WTC collapse dust signature (indicator) as part of the analytical protocols for the sampling and analysis of residual dust in buildings in Lower Manhattan and nearby downwind areas of Brooklyn.

There are, however, sections of the peer review report that delve into discussions that are either beyond the scope of the charge, or that dealt with issues that arose due to EPA's failure to provide relevant information. There were also some reviewers' misconceptions that were not properly vetted during the review process. Prior to finalizing the charge questions to the reviewers the Panel was asked for comment, and Greg Meeker provided two pages of comments that were not passed along to the reviewers. In his comments, Mr. Meeker provided additional technical information on some of the issues that the reviewers found troubling, and provided a framework that might have aided them in

Reprinted with permission of the senior author, Dr. Gregory Meeker, USGS, Denver, Colorado. This report was sent to the EPA on July 28, 2005.

their evaluation and the development of their recommendations (comments attached).

The review of the method and the signature concept in general should have been considered in the context of the Phase I study as proposed by EPA. It has been clear for months that the Phase I study would not only include analysis for the signature component(s) but a selected list of the COPCs as well. Analysis of the Phase I data set, including information about building type, age, use, distance from the WTC site, elevation, wind direction on 9/11, and other factors would have provided the final verdict on the signature concept and its applicability to each of the greater issues of interest. This is a point that was discussed by the WTC Expert Technical Panel as late as July, 2005. It is not at all clear that the signature study was reviewed in that context, nor is it clear that all EPA representatives who formulated the charge questions either understood or clearly articulated that concept. The problem of addressing residual WTC contamination four years after the event is extremely complex and difficult. There is no "off the shelf" method that can address the problem by providing yes or no answers, and no approach to the problem can be totally satisfactory with respect to all of the issues that have been the subject of extensive deliberation by the WTC Expert Technical Panel. The signature concept, as proposed and demonstrated through the signature study, is clearly the best and most appropriate course of action available at this time for determining whether or not elevated concentrations of COPC's actually were derived from the deposited WTC Dust. It is only through additional data collection and analysis, along with further refinement of the method, that the process can be truly evaluated. The signature study was only for the purpose of demonstrating that the process could be started. To the extent that the review actually delayed the implementation of the Phase I study in Lower Manhattan and nearby Brooklyn, some of the time spent was warranted to ensure that interpretable results could be obtained for triggering a cleanup. However, the process went on too long, and the peer review process was flawed for three reasons: 1. the lack of specific information on the deliberations of the WTC expert technical panel, 2. there were no significant adjustments to the charge questions based upon Mr. Meeker's concerns and 3. there was no contact with members of the WTC Expert Technical Panel to address specific technical issues.

SUMMARY ADDRESSING PRIMARY DIFFICULTIES IDENTIFIED BY PEER REVIEWERS

Scatter of Data and Exclusion of Data from Three Laboratories

The peer reviewers of the Final Signature Study Report generally agreed that the scatter in the data and exclusion of three laboratories was a serious problem and required further study to explain the results. Several suggestions were made as to the cause of the scatter in the data. The reviewers focused primarily on possible problems with sample preparation, specifically the drop mount and one minute settling. The WTC Expert Technical Panel's signature sub-group agrees that the data show more scatter than desirable. However, given the nature of the material analyzed, and the time available for method development, we believe that scatter in the data is certainly within acceptable limits for the intended purpose of the Phase I survey, as described below in the discussion. Even the data from the laboratories that were not used by EPA in the final analysis show the same trends as the other laboratories, suggesting that there was a problem with inconsistent sample preparation between laboratories. This is a point that could have been addressed in the selection criteria for the laboratories to analyze the Phase I samples. Further, the peer review group almost entirely overlooked two other possible causes for the scatter in the data; (1) the possibility that the "standard" samples sent to the laboratories were not entirely homogeneous from jar to jar; and (2) the possibility that the "standards" may not have been removed from the jars in a way to insure homogeneity between aliquots. Finally, we believe that there may have been significant problems with the 4 Albany spiking material. In retrospect, it probably should not have been used as a pure spiking material for this study. The data derived from the 4 Albany spiked samples should not be considered in the final analysis unless the material used for spiking is clearly demonstrated to represent pure, undiluted WTC Dust.

Failure of the Study to Consider Other Analytical Methods

Since September of 2001, extensive analytical work has been done on WTC Dust by numerous laboratories using a wide variety of analytical techniques. There is a wealth of data in the literature on trace element chemistry, major element chemistry, organic chemistry, optical properties, unique and/or unusual components and more. In addition, several of the panel members have been personally involved in numerous studies involving a wide variety of analytical techniques and methods including XRD, XRF, ICP-MS, infrared spectroscopy, etc. All of these results were considered in the deliberations of the WTC Expert Technical Panel regarding possible signature components and methods to analyze the components. This information was not effectively used by the peer reviewers, nor was any of the members of the signature subgroup contacted to provide informational clarifications during the peer review panel's conference calls. This was a deficiency in the peer review process. Why bother to have knowledgeable experts on the WTC Expert Technical Panel if you do not use their technical expertise effectively in the evaluation of the nature of this complex mixture?

Concerns Regarding False Positives

It has always been recognized that implementation of this method will result in some number of false positives. This is clearly outlined in the method, the referenced literature, and in the WTC Expert Technical Panel discussions. The question is, given a realistic background level, is the number of false positives that are likely to occur acceptable? Based on the signature study, we believe the answer will be yes. In addition, since these will be false positives, and if the selected COPC's go above the recommended Phase I guidelines, the units could be considered for clean up. The final answer to this question, however, cannot be determined until additional data are collected and analyzed during the Phase I study.

Failure of the Study to Thoroughly Analyze the Chemistry (Fe Content) of Slag Wool and Rely Only on Published Values

This criticism comes from an apparent misconception, by most of the reviewers, that the chemical composition of the WTC slag wool was determined by simply using values in the TIMA literature. This statement is totally incorrect. The WTC slag wool composition was determined by considerable analytical work by quantitative electron probe microanalysis of polished samples at the USGS. Only after the composition of WTC MMVF was accurately determined was TIMA and other industry literature consulted to determine the appropriate glass fiber industry name for each chemical fiber type found in WTC Dust. We think that both the peer reviewers and the members of the WTC Expert Technical Panel were done a disservice by the lack of meaningful discussion between members of the two groups prior to completing the peer review group's report.

Concerns Regarding Persistence of Slag Wool

We believe this is a valid concern and should be evaluated, although we believe that, in indoor environments, the amount of slag wool would not be significantly reduced unless the material sat in water for an extended period of time. Tests to address this question could be conducted quickly for minimal expense.

Use of Outdoor Dust in the USGS Spiking Material

It would have been preferable to use all indoor dust in the USGS spiking material. However, the amount of material needed simply was not available and is not available. The USGS spiking material did contain a substantial amount of indoor dust (i.e. unaffected by rain), and this should have been elucidated in EPA's final report. Furthermore, the relative abundances of slag wool in indoor and outdoor dust was not that different, and was primarily due to the higher gypsum abundance in the indoor samples. It is important to recognize that the relative proportion of MMVF types was consistent between outdoor and indoor dust.

Extrapolation to Fibers per Gram

Extrapolation to fibers per gram was done in order to report consistent units between laboratories. The units could have, and probably should have, been fibers per milligram. The use of fibers per gram was not intended to give a false impression of better counting statistics. It was, however, pointed out by the WTC Expert Technical Panel to EPA, prior to the peer review, that actual fiber numbers should be reported with the data to demonstrate actual counting statistics. It should also be pointed out that in most peer reviewed methods for dust and particle analysis actual fiber numbers are adjusted to common units such as fibers per cm^2 and, in doing so, the fiber numbers are significantly elevated over the number actually counted.

Measurement of Fiber Dimensions

Measurement of fiber dimensions was made during the study in order to determine mass of slag wool in each sample. However, it was determined that fiber number (fibers/gram) provided better data. The majority of fibers in the dust are in the 1–10 µm diameter range. A single long 20 µm diameter fiber can significantly affect the calculated mass determination, particularly with low total fiber numbers.

Use of PLM to Identify Slag Wool

PLM was employed at the beginning of the study for fiber counting. If a liaison had been arranged between the two panels, this point could have been easily clarified to the peer reviewers. These data should have been presented in the report. The analyses showed that the PLM results were no better than results obtained using SEM, and the analysis did not require substantially less time. It is true that determining the precise refractive index for slag wool in order to distinguish it from rock wool was not done. However, the amount of rock wool in WTC Dust is minimal (1–3% total MMVF), and the inability to distinguish it from slag wool by not using the exact refractive index would not have made a significant difference in the results. The more abundant soda-lime glass is easily distinguishable from slag wool using a single index oil of 1.605 or 1.550. Other fiber types were also observed in background samples, and those fiber types required x-ray microanalysis for identification.

DISCUSSION

The WTC Expert Technical Panel has, over the last 15 months, generally endorsed the concept of working toward developing a signature for identification of residual WTC Dust contamination in indoor and possibly some outdoor spaces. Through the efforts of numerous individuals within the US Environmental Protection Agency, the United States Geological Survey, the Environmental & Occupational Health Sciences Institute of UMDNJ/Rutgers University, New York University Nelson Institute of Environmental Medicine, University of North Carolina, the city of New York, and other agencies and organizations, a proposal was formulated to use slag wool, along with other dust components, possibly gypsum and concrete particles, as signature or screening components to help identify residual contamination from the dust generated during the collapse of the WTC. A point that was missed by the Peer Review Panel was that it was always thought by the WTC Expert Technical Panel that the method would be used in conjunction with standard methods for the analysis of other selected COPCs, including asbestos, lead, and PAHs on the same samples.

The signature concept was formulated concurrently with a much-revised Phase 1 sampling and analysis plan developed by EPA. During this process several alternatives for a signature were seriously considered and eventually rejected for a variety of reasons. Records of the discussions regarding these alternatives as primary signature components are in the Panel Meeting summaries and other available documents. Alternative signature components that were discussed include combinations of trace elements (e.g., antimony and chlorine), asbestos, PAHs, iron spheres, and others. Even a proposal to use the behavior of microorganisms exposed to WTC Dust was brought to the attention of the WTC Expert Technical Panel.

The proposal to use slag wool as the primary signature component was judged by many on the WTC Expert Technical Panel and others to be sound for several reasons. The reasons include the following: (1) slag wool was found as a major component (10–60%) of *all* pure bulk WTC Dust samples collected shortly after September 11, 2001 and these analyses were completed by numerous groups; (2) slag wool is durable and likely to be persistent in undisturbed spaces for years; and (3) slag wool is easily identifiable by available analytical techniques at low concentrations.

It was also recognized, from the beginning, that slag wool is a common building material and is likely to be present at relatively high concentrations in some areas of interest, thereby potentially generating false positives. It was also recognized that the relative abundance of slag wool in residual WTC Dust could be affected by parameters such as distance from the source, elevation, and ability to penetrate into indoor spaces through small openings. All of these issues were discussed extensively by the WTC Expert Technical Panel on numerous occasions.

Note: The actual robustness of this approach would have already been on the road to testing, if the EPA Phase I plan had been implemented after the July, 2005 meeting; concurrent with the Signature Peer Review. It is now November, and we are no closer to implementing a Phase I plan for Southern Manhattan and nearby Brooklyn.

In addition to the above, another issue raised and discussed later in the process of signature development by WTC Expert Technical Panel members and others involved in its deliberations was how a signature was to be used. There are many aspects to this question, however, with respect to the proposed Phase I sampling and analysis plan it became clear to the scientists involved that any proposed WTC signature would require continuous and extensive evaluation *during* the Phase I study. The usefulness of a signature would need to be iteratively evaluated for a variety of applications and factors including, but not limited to, identification of specific impacted units and buildings, the geographical extent of contamination, proximity to new construction, and types of areas to which the signature might be applied, e.g., apartments, offices, ages of structures, etc. The peer reviewed signature validation study, as designed, should have been used to answer one primary question; to what extent could a set of qualified laboratories identify and differentiate samples of background dust from samples of that same background dust spiked with pure WTC collapse dust. In addition, the implementation of the Phase I study would have also helped define an initial average background level of slag wool for the impacted areas, and would have provided information about levels of quantification that could be expected. All other scientific questions relating to the application of the signature to issues being addressed by the WTC Expert Technical Panel and EPA would properly be addressed by analysis of data acquired during the Phase I study.

The peer reviewers were given 9 charge questions. Question 1 deals with the design of the study design. All of the peer reviewers agreed that the study was properly designed and, for the most part, properly executed. However, as noted above, we believe the EPA signature study was not evaluated in the proper context.

Question 2 asked if the reviewers believed that the study demonstrated that slag wool could be used as the signature component. The reported consensus opinion of the peer review committee stated "the proposed method has not demonstrated the utility of slag wool as a successful signature constituent." It was also stated in the review that "the information was insufficient to reject slag wool as a signature." These consensus opinions were derived with four of the reviewers answering YES to question 2, 1 reviewer answering NO, and 1 reviewer answering MIGHT BE. With the four YES votes came several questions and reservations. The other reviewers cited additional problems with the study. Many of those reservations were addressed above.

We, as members of the WTC Expert Technical Panel signature subgroup believe that many of the problems with question 2 that were cited by the peer reviewers can be effectively addressed by EPA. For example, if the data for the analyses are examined, as in figure C.1, it is clear that nearly the same trend was observed by all laboratories even though absolute values varied considerably. The data generally show low background levels, with the exception of one high value (C1-RTP), and proportionally higher values for the USGS spiked samples. This suggests variations in the preparation of samples by individual laboratories, but general consistency within laboratories. The 4 Albany samples should not be considered as samples spiked with pure WTC Dust. For whatever reason, the spiking material appears to have been diluted by a factor of 5 to 10 prior to preparation of the spiked test samples. Therefore, these samples appear to have undergone a double dilution whereby the actual concentrations of WTC Dust would have been 1%, 0.5%, and 0.1%, rather than the expected 10, 5 and 1%.

Regarding question 5, it should be noted that the full details of how the "standards" were prepared were not provided to the reviewers. The data in . . . the Final Report on the World Trade Center (WTC) Dust Screening Method Study were obtained on aliquots taken from the blended samples prior to those samples being sent to EPA to be split into aliquots that were then sent to the 8 laboratories involved in the study. The details of how and where these primary blended samples were split were never provided in the final report. The USGS has ex-

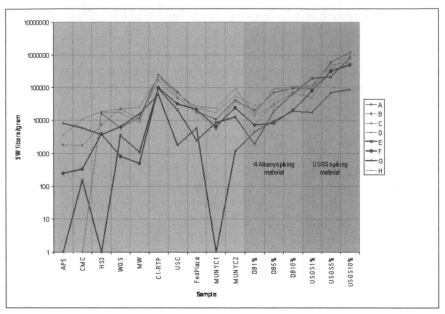

Figure C.1. Variability in slag wool (SW) fiber counts among samples analyzed by different laboratories for WTC dust. Reprinted with Permission of Dr. Gregory Meeker, USGS, Denver, Colorado.

tensive experience in preparation of homogeneous reference materials. It is extremely difficult to blend such materials particularly when the materials are relatively coarse grained and contain dissimilar particle types such as fine granular dust and fibers. It is equally difficult to split a large amount of blended sample into many uniform individual containers ensuring that each container contains precisely the same material. Normally when a reference material is produced, random containers of the material are tested for homogeneity between and within containers prior to distribution. We wonder if some of the scatter observed in the data may be a result of the "standards" that were sent to the laboratories not being identical or the laboratories not sampling from the bottle in a way to insure identical aliquots.

In summary, we believe that many of the objections to the EPA signature validation study outlined by the peer reviewers come as a result of a failure to provide the reviewers with information regarding how the method was to be used in the Phase I sampling plan. We also believe that some assumptions that were made by the reviewers should have been verified through questions provided to the EPA scientists or through them to the WTC Expert Technical Panel. Finally, we have some concerns as to how the EPA signature validation study was conducted with respect to splitting of spiked samples and also to the use of the 4 Albany spiking material. Despite these specific concerns, we believe that the EPA signature validation study has sufficiently demonstrated that laboratories can distinguish between properly spiked background and unspiked background with an acceptable false positive rate. We therefore urge EPA to continue with plans to use the WTC signature as outlined in the Phase I study for Lower Manhattan and nearby Brooklyn.

COMMENTS ON DRAFT CHARGE QUESTIONS FROM GREG MEEKER

Purpose of the Signature Study

The study only addresses to what extent laboratories can identify spiked samples in a series of "background" samples using a pre-selected set of collapse dust components, i.e., proposed signature components. The study also provides preliminary levels for background for the proposed components. If validated, this method will be used to address two questions. These two questions are not adequately discussed in the Draft Charge Questions. These questions are:

1. What is the geographical extent of dust contamination from the collapse and later transport of materials? A primary goal of this process is to use a WTC Dust signature to help evaluate the geographical extent of collapse dust contamination. It should be recognized that after four years this task can only be done as an iterative process whereby newly collected samples from Phase 1 (in general proceeding outward from the WTC site) are analyzed, the data are continually evaluated, and necessary adjustments are made to the methods and final data objectives. The purpose of the signature evaluation study was to evaluate the analytical capabilities of the laboratories in detecting the components using the methods developed. The study did not address how the dust components were

distributed during and after the collapse. It should also be recognized that there *may* be conditions such as distance from the WTC site or elevation that will limit the extent to which the method can provide an estimate for the extent of contamination, but that the method may work very well within a specifically defined area that has yet to be totally determined. The area sampled by USGS [see report by Clark et al., referenced in the bibliography] provides an area of "high confidence" for application of the signature because these samples all contained similar concentrations of slag wool (approximately 10 to 50% with an average value of about 30%). The data from Governors Island presented by EPA at the July 12 Panel meeting strongly suggests that the slag wool signature will be valid to a much greater distance. Given the above background information and caveats, are the results of the study robust enough so that contract laboratories, possibly with assistance of government laboratories, can proceed to use the method to collect data for EPA to begin the iterative process of evaluating the extent of contamination using samples collected during Phase 1?

2. Given that a sampling "unit" is within an area *demonstrated to have been impacted by the collapse dust,* and that "unit" shows elevated COPCs, can any of the proposed signature components be used to demonstrate that the "unit" has or, more likely, *has not* been impacted by WTC Dust? It is a given that there will be some number of false positives (acceptable number or percent yet to be determined, the number of false positives will increase as the number selected for a lower cutoff decreases) but there should not be a significant number of false negatives.

These two questions are different and the samples required to answer them may not necessarily be entirely the same. There may be sampling opportunities that could help address question 1 that will not fit the proposed statistical sampling plan designed largely for question 2. If these sampling opportunities arise they should be taken advantage of and not overlooked because they do not fit a predetermined sampling plan.

Additional Comments

- It must be made clear that the pure DB spiking material contained only about 2% slag wool. This number is significantly less than what was found in the 37 USGS samples and other bulk samples collected by Lioy and other groups (~30% slag wool). The reason for this is not known at this time. The analysis data of the two spiking materials should be included with the data. The DB spiking material is suspect as a "pure" WTC Dust in that the major component ratios do not agree with ANY of the other known and well characterized bulk samples. However, the data from the samples spiked with the DB material still appears to demonstrate that the analysis method can be extended to levels well below what appears to be the average background determined in this study. The USGS spiking material was a composite of several of the bulk WTC Dust samples shown on the attached map. This information should be provided for the peer review.
- The measured number of fibers, not the calculated fibers per gram, determines *the maximum* accuracy these measurements can have. These numbers must be included with the data.
- The full details of sample collection and preparation including production of reference spiking material, background dust and spiked samples must be included for the peer review. Analytical results on these materials along with laboratory calibration data on the supplied reference material must also be included. This information is a critical part of the overall study.
- Details on all of the methods actually used for analysis of the various components by each of the different laboratories should be included (e.g., software or method actually used to calculate gypsum and concrete).
- Exactly what background documents will be included in the package sent to the reviewers?
- Will there be a mechanism for the peer reviewers to ask for and obtain additional information if they feel the information provided is inadequate to properly evaluate the study?
- Q1 (c) should read "They are *sufficiently* homogeneous in WTC Dust." None of the WTC Dust samples are truly homogeneous.

- Where are the references for the documents mentioned in the Charge Questions? It is not clear which documents are being referred to.
- The term "elements of concrete" should be replaced by "phases compatible with concrete." How did each laboratory interpret "Ca-rich particle" for the concrete analysis.
- Will plots of the data be provided to the reviewers as presented at the July 12 Panel meeting?
- The idea that some level of false positives is acceptable should be clearly stated.
- Relative levels of major, minor and trace components of bulk WTC Dust are known. This data should be provided so that the reviewers can judge the range of levels of components, including COPCs that will be present at the dilutions discussed in the signature study.
- Why did only one laboratory analyze the "28 different" background samples and how will these analyses be confirmed?
- In the data tables, do the Calcium-rich values include gypsum or has the gypsum been subtracted?

GLOSSARY

Absorption—the act or process of absorbing or the condition of being absorbed. A route by which substances can enter the body through the skin. In chemistry, absorption of particles of gas or liquid in liquid or solid material. In pharmacokinetics, absorption of drugs in the body.

Adsorption—the formation of a gas or liquid film on a solid surface.

AEGL—Acute Exposure Guidelines. Guidelines established by the U.S. EPA to provide guidance on the levels above which highly toxic substances could cause acute health effects ranging from irritation to death.

Asbestos—any of several minerals (as chrysotile) that readily separate into long flexible fibers, that cause asbestosis and have been implicated as causes of certain cancers, and that have been used especially formerly as fireproof insulating materials.

Coarse particles—particles between 2.5 and 10 µm in diameter. They penetrate and deposit effectively in the upper airways dropped air of the respiratory system.

Epidemiology—a branch of medical science that deals with the incidence, distribution, and control of disease in a population. The sum of the factors controlling the presence or absence of a disease or pathogen.

Exposure—contact with a material at a boundary (nose, skin, or mouth) between the human and the environment at a concentration present in the environment over an interval of time.

Fine particles—particles with a diameter of less than 2.5 µm. They can penetrate effectively and deposit deep in the respiratory system.

Fingerprint—a unique pattern indicating the presence of a particular molecule, based on specialized analytic techniques such as mass- or X-ray spectroscopy, used to identify a pollutant, drug, contaminant, or other chemical in a test sample.

Morphology—the biological study of the form and structure of living organisms. The structure and form of an organism excluding its functions.

NAAQS—National Ambient Air Quality Standard set by the U.S. Environmental Protection Agency. There are six: particles, ozone, sulfur dioxide, nitrogen dioxide, lead, and carbon monoxide.

PAL—Provisional Advisory Level. PALs are a set of exposure values used to inform risk-based decision-making during a response to environmental contamination by hazardous chemicals. They are for exposures to chemicals by the general public, including susceptible and sensitive people.

pH—a measure of the acidity or basicity of a solution. It is defined as the cologarithm of the activity of dissolved hydrogen ions (H^+). Hydrogen ion activity coefficients cannot be measured experimentally, so they are based on theoretical calculations. The pH scale is not an absolute scale; it is relative to a set of standard solutions whose pH is established by international agreement.

Pica—from the dictionary, the word pica "comes from the Latin name for magpie, a bird known for its unusual and indiscriminate eating habits. In addition to humans, pica has been observed in other animals, including the chimpanzee."

Plume—an elongated and usually open and mobile column or band (as of smoke, exhaust gases, or blowing snow).

Recovery—the regaining of something lost or taken away (e.g., the body of a deceased person).

Reentry—to return to the locations affected by a catastrophe to assess damage and determine the nest steps to develop a plan for restoration.

Rehabitation—the return of the people to restart operations in a community or building or facility that had been destroyed or damaged during a catastrophe

Rescue—to set free from confinement or danger, implying freeing from imminent danger by prompt or vigorous action.

Respirator—a device worn over the mouth and nose to protect the respiratory tract by filtering out dangerous substances (such as dusts or fumes) from inhaled air. A device for maintaining artificial respiration.

Restoration—bringing equipment or resource to its operational status after repairs or replacement of parts; bringing a community or larger area back to its original full-scale operations after a disaster.

Slag wool—a type of mineral wool made by forcing steam through molten slag; used as thermal insulation.

Semivolatile—a term describing chemicals that break down very slowly in the environment.

Supercoarse particles—particles greater than 10 μm in diameter. They deposit in the upper airways in the respiratory system and are effectively removed by the nose.

BIBLIOGRAPHY

ATSDR and NYC Department of Health and Mental Hygiene. *Final Technical Report of the Public Health Investigation to Assess Potential Exposures to Airborne and Settled Surface Dust in Residential Areas of Lower Manhattan, World Trade Center Environmental Assessment Working Group*. New York, September 2002.

Bai, Zhipeng, Lih-Ming Yiin, David Q. Rich, John L. Adgate, Peter J. Ashley, Paul J. Lioy, George G. Rhoads, and Junfeng Zhang. Field evaluation and comparison of five methods of sampling lead dust on carpets. *American Industrial Hygiene Association Journal* 64, no. 4 (2003): 528–532.

Bates, David V. *Environmental Health Risks and Public Policy*. Vancouver: University of British Columbia Press, 1994.

———. *Five Minutes into the "Eroica."* N.p.: Friesens, 1997.

Chemical and Engineering News. Terrorism's legacy. January 9, 2006, pp. 36–40.

Clark, Roger N., Robert O. Green, Gregg A. Swayze, Greg Meeker, Steve Sutley, Todd M. Hoefen, K. Eric Livo et al. Environmental studies of the World Trade Center area after the September 11, 2001 attack. USGS Open-File Report 2001-0429, November 27, 2001. http://pubs.usgs.gov/of/2001/ofr-01-0429/.

DePalma, Anthony. What happened to that cloud of dust? *New York Times*, November 2, 2005.

Edelman, Philip, John Osterioh, James Pirkle, Sam P. Caudill, James Grainger, Robert Jones, Ben Blout, et al. Biomonitoring of chemical exposure among New York City firefighters responding to the World Trade Center fire and collapse. *Environmental Health Perspectives* 111, no. 16 (2003): 1906–1911.

Farfel, Mark, Laura Digrande, Robert Brackbill, Angela Prann, James Cone, Stephen Friedman, Deborah Walker, et al. An overview of 9/11 experiences and respiratory and mental health conditions among World Trade Center Health Registry enrollees. *Journal of Urban Health* 85, no. 6 (2008), PMID: 18785012.

Fireman, Elizabeth M., Yehuda Lerman, Eliezer Ganor, Joel Greif, Sharon Fireman-Shoresh, Paul J. Lioy, Gisela I. Banauch, Michael Weiden, Kerry J. Kelly, and David J. Prezant. Induced sputum assessment in New York City firefighters exposed to World Trade Center dust. *Environmental Health Perspectives* 112, no. 15 (2004): 1564–1569.

Freeman, Natalie C. G., Paromita Hore, Kathleen Black, Marta Jimenez, Linda Sheldon, Nicolle Tulve, and Paul J. Lioy. Contributions of children's activities to pesticide hand loadings following residential pesticide application. *Journal of Exposure Analysis and Environmental Epidemiology* 15, no. 1 (2005): 81–88.

Georgopoulos, Panos G., Paul Fedele, Pamela Shade, Paul J. Lioy, Michael Hodgson, Atkinson Longmire, Melody Sands, and Mark A. Brown. Hospital response to chemical terrorism: Personal protective equipment, training, and operations planning. *American Journal of Industrial Medicine* 46, no. 5 (2004): 432–445.

Georgopoulos, Panos G., and Paul J. Lioy. From theoretical aspects of human exposure and dose assessment to computational model implementation: The Modeling Environment for Total Risk Studies (MENTOR). *Journal of Toxicology and Environmental Health—Part B, Critical Reviews* 9, no. 6 (2006): 457–483.

Geyh, Alison S., Steven Chillrud, D'Ann L. Williams, Julie Herbstman, J. Morel Symons, Katherine Rees, Barbara Turpin, Ho-Jin Lim, Sung Roul Kim, and Patrick Breysse. Assessing truck driver exposure at the World Trade Center disaster site: Personal and area monitoring for particulate matter and volatile organic compounds during October 2001 and April 2002. *Journal of Occupational and Environmental Hygiene* 2, no. 3 (2005): 179–193.

Gillin, Eric. What did I breathe in on September 11? *Esquire* magazine, April 2007.

Gonzalez, J. *Fallout*. New York: New Press, 2002.

Hore, Paromita, Mark Robson, Natalie Freeman, Junfeng Zhang, Daniel Wartenberg, Haluk Özkaynak, et al. Chlorpyrifos accumulation patterns for child accessible surfaces and objects and urinary metabolite excretion by children for two-weeks after crack-and-crevice application. *Environmental Health Perspectives* 113, no. 2 (February 2005): 211–219.

Ilacqua, Vito, Natalie C. J. Freeman, Jerald Fagliano, and Paul J. Lioy. The historical record of air pollution as defined by attic dust. *Atmospheric Environment* 37, no. 17 (2003): 2379–2389.

Landrigan, Phillip J., Paul J. Lioy, George Thurston, Gertrud Berkowitz, Steven Chillrud, S. Gavett, Panos Georgopoulos, et al. Health and environmental consequences of the World Trade Center. *Environmental Health Perspectives* 112, no. 6 (2004): 731–739.

Lioy, Paul J. Exposure assessment: Utility and application within homeland or public security. Editorial. *Journal of Exposure Analysis and Environmental Epidemiology* 14, no. 6 (2004): 427–428.

Lioy, Paul J., and Panos G. Georgopoulos. The anatomy of the exposures that occurred around the World Trade Center site: 9-11 and beyond. In *Living in a Chemical World Framing the Future in Light of the Past*, vol. 1076, pp. 54–79. New York: New York Academy of Sciences, 2006.

Lioy, Paul J., and Michael Gochfeld. Lessons learned on environmental, occupational, and residential exposures from the attack on the World Trade Center (WTC). *American Journal of Industrial Medicine* 42, no. 6 (2002): 560–565.

Lioy, Paul J., Edo Pellizzari, and David Prezant. The World Trade Center aftermath and its effects on health: Understanding and learning through human exposure science. *Environmental Science & Technology* 40, no. 22 (2006): 6876–6889.

Lioy, Paul J., Fred S. Roberts, Brenden McCluskey, Mary Jean Lioy, Lee Clarke, Audrey Cross, Louise Stanton, William Tepfenhart, and Mary Ellen Ferrara. TOPOFF 3 comments and recommendations by members of New Jersey Universities Consortium for Homeland Security Research. *Journal of Emergency Management* 4, no. 6 (November/December 2006): 41–51.

Lioy, Paul J., Daniel Vallero, Gary Foley, Panos Georgopoulos, John Heiser, Tom Watson, Michael Reynolds, James Daloia, Sai Tong, and Sastry Isukapalli. A personal exposure study employing scripted activities and paths in conjunction with atmospheric releases of perfluorocarbon tracers in Manhattan, New York. *Journal of Exposure Science and Environmental Epidemiology* 17, no. 5 (2007): 409–425.

Lioy, Paul J., Clifford P. Weisel, and Panos G. Georgopoulos. An overview of the environmental conditions and human exposures that occurred post 9-11. In *Urban Aerosols and Their Impacts: Lessons Learned from the World Trade Center Tragedy*, ed. Jeffrey S. Gaffney and Nancy A. Marley, pp. 32–38. ACS Symposium Book. Oxford: Oxford University Press, 2005.

Lioy, Paul J., Clifford P. Weisel, James R. Millette, Steven Eisenreich, Daniel Vallero, John Offenberg, Brian Buckley, et al. Characterization of the dust/smoke aerosol that settled east of the World Trade Center (WTC) in Lower Manhattan after the collapse of the WTC September 11, 2001. *Environmental Health Perspectives* 110, no. 7 (2002): 703–714.

Lorber, Matthew, Herman Gibb, Lester Grant, Joseph Pinto, Joachim Pleil, and David Cleverly. Assessment of inhalation exposures and potential health risks to the general population that resulted from the collapse of the World Trade Center towers. *Risk Analysis* 27, no. 5 (2007): 1203–1221.

Lowers, Heather A., Gregory P. Meeker, Paul J. Lioy, and Morton Lippmann. Summary of the development of a signature for detection of residual dust from collapse of the World Trade Center buildings. *Journal of Exposure Science and Environmental Epidemiology* 19, no. 2 (2008): 325–335.

Maciejczyk, Polina B., Rolf L. Zeisler, Jing-Shiang Hwang, George D. Thurston, and Lung Chi Chen. Characterization of size-fractionated World Trade Center dust and estimation of relative dust contribution to ambient particulate concentrations. In *Urban Aerosols and Their Impacts: Lessons Learned from the World Trade Center Tragedy*, ed. Jeffrey S. Gaffney and Nancy A. Marley, pp. 114–141. American Chemical Society Symposium Series 919. Oxford: Oxford University Press, 2005.

Maremont, Mark, and Jared Sandberg. Weighing risks: Tests say air is safe, but some people feel ill near Ground Zero. *Wall Street Journal*, December 26, 2001.

Meeker, Gregory P., Amy M. Bern, Heather A. Lowers, and Isabelle K. Brownfield. *Determination of a Diagnostic Signature for World Trade Center Dust Using Scanning Electron Microscopy Point Counting Techniques*. Denver: U.S. Geological Survey, 2005.

Millette, James R., Randy Boltin, Pronda Few, and William Turner Jr. Microscopical studies of the World Trade Center disaster dust particles. *Microscope* 50, no. 1 (2002): 29–35.

Millette, James R., Paul J. Lioy, John Wietfeld, Thomas J. Hopen, Mark Gipp, Timothy Padden, Craig Singsank, and Jacqueline Lepow. A microscopical study of the general composition of household dirt. *Microscope* 51, no. 4 (2004): 201–207.

Morgan, Daniel L., Herman C. Price, Michael P. Moorman, Andrew Suttie, and Paul J. Lioy. Evaluation of World Trade Center dust pulmonary toxicity after intratracheal instillation of total particulate. Unpublished report.

Offenberg, John H., Steven J. Eisenreich, Lung Chi Chen, Mitch D. Cohen, Glen Chee, Colette Prophete, Clifford Weisel, and Paul J. Lioy. Persistent organic pollutants in the dusts that settled across Lower Manhattan after 11 September 2001. *Environmental Science and Technology* 37, no. 3 (2003): 502–508.

Offenberg, John H., Steven J. Eisenreich, Cari L. Gigliotti, Lung Chi Chen, Mitch D. Cohen, Glenn Chee, Colette M. Prophete, et al. Persistent organic pollutants in the dusts that settled across Lower Manhattan after 11 September 2001. In *Urban Aerosols and Their Impacts: Lessons Learned from the World Trade Center Tragedy*, ed. Jeffrey S. Gaffney and Nancy A. Marley, pp. 103–113. American Chemical Society Symposium Series 919. Oxford: Oxford University Press, 2005.

Offenberg, John, Steven. J. Eisenreich, Caril Gigliotti, Lung Chi Chen, Judy Q. Xiong, Chunli Quan, Xiaopeng Lou, et al. Persistent organic pollutants in dusts that settled indoors in Lower Manhattan after 11 September 2001. *Journal of Exposure Analysis and Environmental Epidemiology* 14, no. 2 (2004): 164–172.

Prezant, D. J., M. Weiden, G. I. Banauch, G. McGuinness, W. M. Rom, T. K. Aldrch, and K. J. Kelly. Cough and bronchial responsiveness in firefighters at the World Trade Center site. *The New England Journal of Medicine* 347, no. 11 (2002): 806–815.

Revkin, Andrew C. After attacks, studies of dust and its effects. *New York Times*, October 16, 2001.

Skloot, Gwen S., Clyde B. Schechter, Robin Herbert, Jacqueline M. Moline, Stephen M. Levin, Laura E. Crowley, Benjamin J. Luft, Iris G. Udasin, and Paul L. Enright. Longitudinal assessment of spirometry in the World Trade Center medical monitoring program. *CHEST* 135, no. 2 (2009): 492–498.

Swayze, Gregg A., Roger N. Clark, Steve Sutley, Todd M. Hoefen, Geoff S. Plumlee, Greg P. Meeker, Isabelle K. Brownfield, K. Eric Livo, and Laurie C. Morath. Spectroscopic and X-ray diffraction analyses of asbestos in the World Trade Center dust. In *Urban Aerosols and Their Impact: Lessons Learned from the World Trade Center Tragedy*, ed. Jeffrey S. Gaffney and Nancy A. Marley, pp. 40–65. American Chemical Society Symposium Series 919. Oxford: Oxford University Press, 2005.

U.S. Environmental Protection Agency. *Committee of the World Trade Center Indoor Air Task Group, World Trade Center Indoor Environmental Assessment: Selecting Contaminants of Potential Concern and Setting Health Based Benchmarks, EPA Region II, NYC, NY, May 2003.* Washington, DC: Author, 2003.

———. *An Inhalation Exposure and Risk Assessment of Ambient Air Pollution from the World Trade Center Disaster.* National Center for Environmental Assessment, ORD, A DRAFT. Washington, DC: Author, December 2004.

———. *Mapping of the Spatial Extent of the Ground Dust/Debris from the Collapse of the World Trade Center Building.* ORD, NERL, Appendix A, NYC, NY. December 2005. www.epa.gov/wtc/panel/pdfs/WTC5_WTC_Report_Figuresonly_December 2005.pdf.

———. Panel meeting 06-22-2004. WTC Expert Technical Review Panel. June 22, 2004. http://www.epa.gov/wtc/panel/meeting-20040622.html.

————. *Summary Report of the USEPA Technical Peer Review Meeting on the Draft Document Entitled: Exposure and Human Health Evaluation of Airborne Pollution from the World Trade Center Disaster.* EPA/600-R-03/142, ORD EPA. Washington, DC: Author: December, 2003.

————. Whitman details ongoing agency efforts to monitor disaster sites, contribute to cleanup efforts. EPA Response to September 11, press release. September 18, 2001.

U.S. EPA Region II. *EPA's Response to the World Trade Center Collapse: Challenges, Successes, and Areas for Improvement.* GAO Report 2003-P-00012. Washington, DC: Author, August 21, 2003.

————. *Interim Final WTC Residential Confirmation Cleaning Study.* Vol. I and II, USEPA Reg. II Rand RO, New York 2003.

————. *World Trade Center: EPA's Most Recent Test and Clean Program Raises Concerns That Need to Be Addressed to Better Prepare for Indoor Contamination Following Disasters.* GAO-07-1091. Washington, DC: Author, September 2007.

————. *World Trade Center Background Study Report, Interim Final.* IAG No. EMW-2002-IA-0127, EPA Region II, NYC, NY, April 2003.

————. *World Trade Center Residential Dust Cleanup Program, Final Report.* EPA Region II, NYC, NY, December 2005. U.S. GAO. *Emergency Management: Observations on DHS's Preparedness for Catastrophic Disasters.* GAO-08-868T. Washington, DC: Author, June 11, 2008.

Webber, Mayris P., Jackson Gustave, Roy Lee, Justin K. Niles, Kerry Kelly, Hillel W. Cohen, and David J. Prezant. Trends in respiratory symptoms of firefighters exposed to the World Trade Center disaster: 2001–2005. *Environmental Health Perspectives* 117, no. 6 (2009): 975–980.

Wheeler, Katherine, Wendy McKelvey, Lorna Thorpe, Megan Perrin, James Cone, Daniel Kass, Mark Farfel, Pauline Thomas, and Robert Blackbill. Asthma diagnosed after September 11, 2001, among rescue and recovery workers: Findings from the World Trade Center Health Registry." *Environmental Health Perspectives* 115, no. 11 (2007): 1584–1590.

Whitman, Christine Todd. The EPA was right. Op-ed. *Star-Ledger*, September 14, 2003.

Wolff, Mary S., Susan L. Teitelbaum, Paul J. Lioy, Regina M. Santella, Richard Y. Wang, Robert L. Jones, Kathleen L. Caldwell, et al.. Exposures among pregnant women near the World Trade Center site on 9/11. *Environmental Health Perspectives* 113, no. 6 (2005): 739–748.

Yiin, Lih Ming, Paul J. Lioy, and George G. Rhoads. Impact of home carpets on childhood lead intervention study. *Environmental Research* 92, no. 2 (2004): 161–165.

Yiin, Lih-Ming, James R. Millette, Alan Vette, Vito Ilacqua, Chuni Quan, John Gorczynski, Michaela Kendall, et al. Comparisons of the dust/smoke particulate that settled inside the surrounding buildings and outside on the streets of southern New York City after the collapse of the World Trade Center, 11 September 2001. *Journal of the Air and Waste Management Association* 54 (2004): 515–528.

Yiin, Lih-Ming, George G. Rhoads, David Q. Rich, Jungeng Zhang, Zhipeng Bai, John L. Adgate, Peter J. Ashley, and Paul J. Lioy. Comparison of techniques to reduce residential lead dust in carpet and on upholstery: The New Jersey assessment of cleaning techniques (NJ ACT) trial. *Environmental Health Perspectives* 110, no. 12 (2002): 1233–1237.

Specific References on the Health Effects of Air Pollution

Detels, Roger, James McEwen, Robert Beaglehole, and Heizo Tanaka, eds. *Oxford Textbook of Public Health*. Vols. 1–3. 5th ed. Oxford: Oxford University Press, 2009.

Holgate, Steven T., Hillel S. Koren, Jonathan M. Samet, and Robert L. Maynard. *Air Pollution and Health*. New York: Academic Press, 1999.

Swift, David L., and W. Michael Foster. *Air Pollutants and the Respiratory Tract*. New York: Decker, 1999.

INDEX

ABOUT THE AUTHOR

Paul J. Lioy is a professor and vice chair, Department of Environmental and Occupational Medicine at UMDNJ–Robert Wood Johnson Medical School. He is deputy director for government relations and director of exposure science at the Environmental and Occupational Health Sciences Institute of Rutgers University and RWJMS–UMDNJ. He received a BA from Montclair State College, NJ, an MS from Auburn University, AL, and a Ph.D. from Rutgers University, NJ. Dr. Lioy received the International Society of Exposure Science (ISES) Jerome Wesolowski Award for Lifetime Achievement in Exposure in 1998, the Frank Chambers Award for lifetime achievement in Air Pollution from the Air and Waste Management Association in 2003, and the Rutgers University Graduate School's Distinguished Alumnus Award in mathematics, engineering, and physical sciences in 2008. He received a National Conservation Award from the Daughters of the American Revolution, and their founders trustees, Ellen Hardin Walworth Medal for Patriotism in 2009. He has served on the Science Advisory Board, U.S. EPA, and the National Research Council, Board of Toxicology and Environmental Studies. He is a founder and past-president of the ISES. Dr. Lioy is an associate editor of the journal *Environmental Health Perspectives*, and of the *Journal of Exposure Science and Environmental Epidemiology*. He has published over 245 scientific papers and other publications.

The Honorable Tom Kean served as governor of the State of New Jersey from 1982 to 1990, and during his term was rated as one of the nation's most effective leaders by Newsweek magazine. In 2002 he was appointed by President George W. Bush as chairman of the 9/11 Commission. The commission's report, released in July of 2004, became a national bestseller, and its recommendations resulted in the largest intelligence reform in the nation's history. Kean was president of Drew University, Madison, New Jersey, from 1990 through 2005, stressing teaching, creative use of technology in the liberal arts, and international education. Currently he serves on numerous boards, and has a regular column with the *Star-Ledger*, New Jersey.